*1991*

W9-ACE-869

# TV News Ethics

# ELECTRONIC MEDIA GUIDES

# TV News Ethics

Marilyn J. Matelski, Ph.D.
Boston College

**Focal Press**
Boston   London

Focal Press is an imprint of Butterworth–Heinemann.

Recognizing the importance of preserving what has been written, it is the policy of Butterworth–
Heinemann to have the books it publishes printed on acid-free paper, and we exert our best efforts
to that end.

**Library of Congress Cataloging-in-Publication Data**
Matelski, Marilyn J. 1950-
   TV news ethics / by Marilyn J. Matelski.
         p.         cm.—(Electronic media guide)
   Includes bibliographical references.
   ISBN 0-240-80089-3 (paperback)
   1. Television broadcasting of news—United States.
   2. Journalistic ethics—United States.  3. Journalism—United   States—Objectivity.
   I. Title. II. Series.
   PN4888.E8M38   1991
   174'.9097—dc20                                                      90-40736

**British Library Cataloguing in Publication Data**
Matelski, Marilyn J. , 1950-
   TV news ethics.—(Electronic media guide)

   1. Mass media. Ethical aspects
   I. Title. II. Series.
   174.9097

   ISBN 0-240-80089-3

Butterworth–Heinemann
80 Montvale Avenue
Stoneham, MA 02180

10   9   8   7   6   5   4   3   2   1

Printed in the United States of America

*For Mom and Dad—the most ethical people I know*

# Contents

# Acknowledgments

Putting a book together is never an easy task, but writing on television ethics has been one of the most difficult—and rewarding—challenges I have ever attempted. I have not, however, completed this work without some invaluable assistance from both friends and colleagues. First of all, to Dr. Nancy L. Street, who helped me to develop a clear perspective on the problem—and to keep it consistent throughout the book. Secondly, to Charles E. Morris III, a person with great writing and research skills, and someone who always performed both large and small tasks for me with complete professionalism and good nature. Next, to the following TV newspeople, who provided me with invaluable information and candor: Jack Durant (WKTV); Alan Griggs (WSMV); Steve Grund (KWGN); Skip Haley (WBRZ); Milt Jouflas (KSTU); John McKean (KGGM); Dean Mell (KHQ); Henry Mendoza (KBAK); Brian Moore (WVGA); Dick Nelson (KCTV); Marselis Parsons (WCAX); Paul Paolicelli (KPRC); Bob Reichblum (WJLA); David Russell (WHKE); Jane Sachs (WTTE); Donna Skattum (WAYK); Steve Snyder (KDLT); Liz Talbot (WVTV); Kent Taylor (KHAS); and George Wilson (WOFL). Finally, to Karen Speerstra and Phil Sutherland at Focal Press for all their encouragement and support.

# Introduction

The concept of "ethics" is accepted in all societies as the prescribed means of leading one's life. However, specific ethical codes are always dependent upon a culture's religious beliefs, social norms, historical contexts, and economic philosophies. Therefore, when addressing ethical principles, it is important to view the system within which the standard applies. Karl Mannheim refers to such an approach as "relationism"—considering the relation between the philosophy (or universal spirit) of a cultural value and its sociology (or social institution of definition). This point of view differs greatly from the absolutist, who believes that all things *should* be done a certain way, regardless of time or space. It also disputes the relativist's contention that *nothing* is universal; all values are ephemeral and totally dependent upon immediate circumstances.

Furthermore, relationism functions within a "vertical structure." In other words, a given institution should be judged by its ability to function within its mandate, not whether it can be compared cross-culturally. For example, American broadcasters receive a specific mandate from the governmental arm of the FCC. The relationist media critic does not question this directive; instead, he or she questions how well or how badly the institution meets the specific criteria and/or societal goals of the governmental mandate. Thus, the murky waters of absolutism or relativism are not part of either the problem or the solution.

One of the major difficulties which often occurs when academics discuss ethics with broadcast journalists is that scholars are often seen as the "absolutists," emphasizing constantly what should be done by reporters and news directors. TV executives, on the other hand, are often viewed as the "relativists," changeable by time and broadcast whim.

In point of fact, neither approach is productive when evaluating TV news ethics. This is because broadcast journalism cannot ignore its dualistic nature of functioning as a governmental watchdog while making money as a capitalist business.

Radio broadcasting was identified as a private enterprise which utilized public airwaves, according to the 1927 Federal Radio Act. By this very definition, programmers were ordered to serve their audiences responsibly; they were also encouraged to make enough money to stay in business. This dichotomy of purpose is not always apparent; however, when discussing the ethics of electronic journalism, it cannot be minimized nor ignored.

Thus, to better understand the ethical dilemmas facing TV newscasters today, this text reflects a "relationist" perspective, balancing generally accepted American codes of ethics with the special dilemmas confronting commercial television station owners. By assuming this position, the author hopes to encourage more meaningful dialogue between media scholars and broadcast professionals.

With this in mind, it is important to articulate the specific goals and objectives behind *TV News Ethics*. First and foremost, this book serves as a useful tool of ethical analysis for television news directors and new employees in the industry, as well as for mass communication students. Through a construct known as *Potter's Box* (discussed specifically in Chapters 3 and 4), for example, readers can create a philosophical framework and track news decision-making patterns within a station. They can then compare this pattern to one that reflects their own personal and professional ethical standards. This exercise provides a useful perspective on current ethics in electronic journalism. It also recognizes models of behavior rather than focusing on isolated incidents.

In short, this book is intended to stimulate thinking, not necessarily to provide answers for specific ethical dilemmas. As a result, one will not read long discussions about traditional philosophies. Instead, in *TV News Ethics*, one will discover ways to determine (or formulate) the ethical parameters of a station.

Finally, a detailed bibliography can be found at the end of the text. It is designed to provide further referencing on philosophical approaches as well as other ethical perspectives and sources.

# 1

# ▼ Ethics
# ▼ in the TV News Setting

On November 13, 1969, former Vice President Spiro T. Agnew fired a shot heard 'round the world (of media). In response to what he termed "an instant analysis" by network commentators of a Nixon speech, Agnew took the opportunity to attack television news coverage in general. His now famous Iowa address to the Midwest Regional Republican Committee about news commentaries quickly became the cry of the nation.[1] It still seems relevant today.

In his message, Agnew posed some penetrating questions to the American public: Are we demanding enough of our television news presentations? Are the men and women of this medium demanding enough of themselves? Is there a double standard of ethics where electronic journalism is involved?

After the speech, thousands of people wrote the networks and/or phoned in their reactions to Agnew's allegations. The telephone calls were evenly divided in support of television and the vice president; however, letters and telegrams showed overwhelming support for Agnew's position, with ratios ranging from 6:1 to 10:1, depending on the network.[2]

Among his many charges, Agnew accused the electronic media of being little more than a renegade band of self-appointed leaders. According to the vice president, this irresponsible minority of newscasters must have apparently viewed Americans as a flock of sheep.

> *Agnew:* When the President completed his address—an address, incidentally, that he spent weeks in the preparation of—his words and policies were subjected to instant analysis and querulous criticism. The audience, of 70 million Americans gathered to hear the President of the United States, was inherited by a small band of network commentators and self-appointed analysts, the majority of whom expressed in one way or another their hostility to what he had to say.[3]

Further, he suggested that this small band of men were in collusion with each other when deciding how to present the news.

> *Agnew:* The purpose of my remarks tonight is to focus your attention on this little group of men who not only enjoy a right of instant rebuttal to

every Presidential address, but, more importantly, wield a free hand in selecting, presenting, and interpreting the great issues of our nation . . . . We can deduce that these men read the same newspapers. They draw their political and social views from the same sources. Worse, they talk constantly to one another, thereby providing artificial reinforcement to their shared viewpoints.[4]

Finally, Agnew asked the American people to reflect upon their fate if TV news operations were allowed to continue in the same way.

*Agnew:* By way of conclusion, let me say that every elected leader in the United States depends on these men of the media. Whether what I've said to you tonight will be heard and seen at all by the nation is not my decision, it's not your decision, it's their decision . . . . The great networks have dominated America's airwaves for decades. The people are entitled to a full accounting of their stewardship.[5]

Needless to say, broadcast journalists everywhere from New York to Pocatello were enraged by these pseudocharges. The newspeople reacted most heatedly to Agnew's charge of "instant analysis," since they had seen the speech at least an hour or two earlier, and had been briefed by Henry Kissinger beforehand. As for Agnew's comment that newscasters continuously talk to each other, thereby presenting a deceptively uniform interpretation of news events, Av Westin, then executive producer of the "ABC Evening News," most clearly articulated his colleagues' response. He stated: "I'll be damned if I'll compare notes with anyone. My job is to beat the hell out of the networks."[6] Reuven Frank, then NBC News president, further articulated the unfairness he felt in Agnew's remarks:

The Vice President's charge that a "narrow and distorted picture of America" which over-emphasizes violence and lawlessness was just another example of the messengers being blamed for the message. If we hurt them with our coverage, they say we say too much. If we hurt them by not covering, they say we say too little.[7]

Regardless of what position was taken in this controversy, it was clear that the Agnew speech caused extreme consternation for electronic journalists as well as for the rest of the nation. However, probing ethical questions about TV news presentation did not end there. Since 1969, similar overtones have been voiced by other media critics, and their comments seem equally noteworthy. Typical stories about potential ethical news abuses include the following:

- News correspondent Carl Stern once reported on the possibility of a Supreme Court appointment for Senator Robert Byrd. Stern went on to discuss Byrd's academic record at the American University Law School, and spoke of "much publicized charges some years ago—never proved—that congressional employees wrote his term papers." There was no support to

Stern's claim, and Senator Byrd protested the violation of his rights, but NBC never made an effort to retract it. The potential damage to Byrd's political career was evident.[8]

• John Chancellor, while still a news correspondent, once spent weeks on a story about living in Newark, New Jersey (at that time, one of the most decaying cities of the nation). After a homeowners association complained that real estate prices might be affected by the report, NBC chose not to broadcast the news film—and Chancellor was never even acknowledged for all his work.[9]

• During the Vietnam War, a CBS News cameraman persuaded an American soldier to cut off the ear of a dead Viet Cong military man. The cameraman then sent the footage to news correspondent Don Webster, who taped his commentary and shipped the package to New York. It aired on "The CBS Evening News With Walter Cronkite," October 11, 1967.[10]

• On June 20, 1985, all three television networks gave maximum broadcast coverage to a Beirut press conference called by the hijackers of TWA Flight 847. In this conference, 40 American hostages read written statements while armed Shi'ite gunmen stood by. Shortly afterwards, many people asked whether this extensive news coverage actually aided the hostages; perhaps it helped the captors more.[11]

• In 1989, a nationally known broadcast journalist carried a gun inside the U.S. Capitol, just to prove his point that security needed to be tightened up considerably.[12] His investigative reporting technique was subsequently criticized for its lack of responsibility and tabloid flavor.

Despite these admitted errors in ethical judgment, however, electronic journalists and/or news directors have often shown themselves to possess high levels of sensitivity, integrity, and concern for truth.

• In July 1979, Pierre Salinger presented a profile of the famous Spanish bull-fighter, El Cordoba, for "20/20." Cordoba's story was especially poignant, since he was planning to reenter the ring after a long absence from the sport. However, "20/20" co-anchor Hugh Downs, an avid animal rights activist, was obviously disturbed by the story since, to him, it was more like a promotion for the cruelty of bullfighting than a tribute to El Cordoba. Downs writes:

"I know the rationale here," I prodded Pierre on camera at the close of his piece. "The bullring might well be less inhumane than the slaughterhouse, but the bull is outnumbered any way you cut it, and it is very hard to justify."[13]

• During the long Iranian hostage crisis in the late seventies, Walter Cronkite decided to end his telecast each night with a reminder of how many days the victims had been held captive. This effort kept the story alive, even after reporters and cameramen had been banned from Tehran, and the Carter administration had attempted to diffuse the problem.[14]

- In 1986, ABC's Barbara Walters was given an exclusive interview with arms merchant Manucher Ghorbanifar (a key figure in the 1985-1986 Iranian "arms for hostages" plan). After the interview, Ghorbanifar asked Walters to send a private message to then President Reagan. She did. While later criticized as being a participant, rather than a reporter, in her own story, Walters nonetheless believed that this information might have helped the remaining hostages. To her, the story did not end after the segment aired.[15]

- Several years ago, "60 Minutes" produced a detailed expose on the cereal and hair dye industries. This occurred, despite the fact that two of CBS's largest advertisers, General Foods and Clairol, might be offended by the information presented.[16]

- Recently, *TV Guide* revealed that Dan Rather has continued to question CBS's news gathering methods during the 1980s Afghanistan war, despite an October 1989 memo from network executives declaring the matter "officially closed." The actual controversy began with a charge put forth by a *New York Post* reporter, suggesting that some of the Afghan combat footage was falsified. Rather felt that CBS did not investigate the matter as fully as possible, because, as he says, "[News coverage] is an ongoing process."[17]

As illustrated in the instances above, many examples (too numerous to mention here) do exist where TV reporters have made serious ethical errors in judgment. However, there have also been many times where these reporters have proven to be responsible, sensitive, and reasonably objective professionals. Hence, one of the key issues in electronic news gathering seems to be *intentionality*. In other words, can TV journalists really present the most objective news possible at all times? Or are there inherent limitations which will continue to impede the process of information dissemination?

First of all, it is important to make a very basic assumption: all communication possesses some persuasive dimension. It's inherent in the process. This is because each individual tends to see things through his or her own unique systems of beliefs, attitudes, and values. This is also why truth can never be absolute; it can only be probabilistic. Thus, if objectivity is impossible to achieve at an interpersonal level, it is clear to see that the communication process becomes even more complex when broadcasters attempt to distribute information through the mass media.

Bias, or selectivity, occurs in all links of the news chain. First, there is the event, which the reporter (with his or her own beliefs, attitudes, and values) perceives to happen in a certain way. Next, the story goes to film editors and

management, both of whom consciously and subconsciously determine what
will be shown on the air. Higher management, advertisers, government officials, etc., often enter into the conscious decision-making process; personal/
professional beliefs, attitudes, and values have a part in the subconscious selection. From the editing room to the anchorperson's desk, another selective decision is involved—how much time can be used for each segment? The anchors
then reflect their personal biases through vocal tone, nonverbal cues, etc. And
last, but not least, the viewers choose to accept or reject what they deem important according to their own beliefs, attitudes, and values.

Other major limitations to news dissemination include: (1) time constraints; (2) visual aesthetics; (3) the emphasis on entertainment in television;
and (4) economic considerations.

## TIME

In every half hour newscast, there are approximately 22 minutes and 15
seconds of actual newstime; in every hour-long broadcast, the news allotment
approximately doubles. This is not a great amount of time, considering the
material that must be covered. Consequently, brief news headlines have become the accepted norm for television—people usually read a paper or magazine for more in-depth reporting. Walter Cronkite often has alluded to this
point. He says that the electronic media (radio and television) and the press are
complementary; broadcasting gives the headlines, print gives the depth. Unfortunately, however, most people watch *only* the headlines on the screen, and
therefore receive an incomplete picture of what *really* happened (if anyone
knows). Television (or radio) is then blamed for fostering an uninformed, naive
American people.

Another limitation of time is sometimes considered its greatest strength in
televised news—its immediacy. While it is true that no one can release information faster than those in the broadcast industry, one must ask whether this
information is always useful. Tamotsu Shibutani, in *Improvised News: A Sociological Study of Rumor*, traced the transmission of several rumors after the
Kennedy assassination in 1963.[18] Verified news was at a minimum; and
newspeople, trying to stay ahead of fast-moving events, televised every piece
of information they could find. As a result, rumors were so numerous, no one
knew what to believe. With time, of course, most of the false reports were
dispelled—but that didn't help the confusion which had already begun.

Since 1963, the problem of false information for the sake of immediacy
has not improved significantly. Witness, for example, Frank Reynolds' premature emotional breakdown during John Hinkley's assassination attempt on
President Reagan in 1981. Or how about the large amounts of misinformation
during the November 1989 San Francisco earthquake? These instances underscore the fact that presenting news faster does not always mean it's been presented better.

## VISUAL AESTHETICS

Since television is both a sight and sound medium, it is uniquely different from other information sources. At first glance, this may not seem like a limitation; however, as one looks deeper, visual aesthetics appear to be an important factor in TV news criticism. In television, unlike other media, an ongoing sensory competition often occurs between the eye and the ear. Not surprisingly, "the eye" wins out, and the picture supersedes the sound. When carried to its extreme, the idea of pictorial importance may become a preoccupation, with producers concentrating on visual effects, rather than on the news. As a result, some news stories may often be deleted altogether, simply for lack of visual power; other stories may be glorified because of them. Jan R. Costello, a former TV reporter, illustrates this point masterfully in "Exploiting Grief: Restraint and the Right to Know." This article raises many ethical questions about newscasting, some of which surround the distasteful by-products which can occur with the overemphasis of video journalism:

> Why do reporters point the cameras at grieving families, heartbroken widows, and suffering accident victims? Why do photojournalists and television news reporters go to such lengths to capitalize on an unfortunate person's tragic and gory death? Because that is the quick and dirty way to convey emotion, and emotion sells news. People expect to see human drama and misery unfold in their living rooms. Tears, anguish, and pain are what make photographs and television stories powerful.[19]

As Costello suggests, words are often given little credibility when reporters try to relay tragic accidents and human suffering—the pictures are the primary persuaders. Thus, the question arises: Where should our news priorities lie?

## THE EMPHASIS ON ENTERTAINMENT IN TV

Remember the two college students who ate dog food for a week? And how about the car thief who picked up a hitchhiker, only to discover that he was the owner of the car? You guessed it—these stories and more like them have reached the desks of both local and network news anchors.

Despite denials from many broadcast executives, television has, in fact, developed into a total entertainment medium. Any formalized movement to inform, persuade, or educate the public has been largely incidental, according to Av Westin, former executive producer of "The ABC Evening News." This is because viewers don't wish to hear about disheartening events at the end of a long work day. They want to spend a happy, quiet evening at home. More specifically:

> The audience at dinner time wants to know the answers to three very important questions: Is the world safe? Is my hometown and my home safe?

If my wife and children are safe, then what happened in the past 24 hours to make them better off or to amuse them?[20]

Westin's commentary brings up a further point. Television programming, whether informational or entertainment-oriented, depends heavily upon ratings for advertising support. Thus, viewer popularity is an important consideration when determining how to present the evening news.

The worldwide commitment to famine in Ethiopia is a prime example of news presentation being directly relatable to perceived viewer reactions. Michael Ignatieff writes:

> There is little doubt, first of all, that the television coverage of Ethiopia had a remarkable impact upon Western charity. . .TV brought public pressure to bear upon the bureaucratic inertia, logistical stumbling blocks, and ideological excuses that had allowed a long-predicted food crisis to become a disaster. Television helped to institute a direct relation of people to people which cut through bilateral governmental mediations. For a brief moment, it created a new kind of electronic internationalism linking the consciences of the rich and the needs of the poor. As a medium, television dramatically reduced the lag-time between pressure and action, between need and response. Without it, thousands more Ethiopians would have died, as they have died unseen and unlamented by the West in the famines that have ravaged the country nearly every twenty years in this century.

> Yet if this is the case for television's good conscience over the Ethiopian story, there are other more troubling aspects to the gaze it cast on disaster. There is the accusation that TV news ignored food shortages until they acquired the epic visual appeal of famine, and there is the suspicion that the story will drop out of the nightly bulletins when the focus upon horror shifts elsewhere in the world. The medium's gaze is brief, intense, and promiscuous. The shelf life of the moral causes it makes its own is brutally short.[21]

Television, as a commercial industry, has learned to cater to viewers' personal whims. Certain stories with negative appeal are often discarded prematurely, while others, though possibly irrelevant, endure as long as they are popular.

## ECONOMIC CONSIDERATIONS

Relatedly, the news business needs spot buyers (advertisers) to continue its existence; sponsors do not necessarily buy commercial time if they sense that viewers might become sad or depressed. Hence, it can be said that advertising often interferes with two aspects of broadcast news. One deals with the entire show, the other with story content.

In the former case, there seems to be a great preoccupation with ratings. If news doesn't fare well against certain competition, it is liable to be moved to another time period, lose its sponsor, or (at the very least) lose some valuable ad revenue. A station in a major market, for example, can lose as much as $1 million in income for the loss of one mere ratings point.[22] This may not be a pleasant thought, but it is a fact of life. Thus, at the risk of cheapening broadcast journalism, news writers often deal in sensationalism to capture an audience.

As for story content, the measurement of advertiser control in news presentation is often more subtle than the battle for ratings points. News directors may choose not to cover certain stories, for example, simply to avoid offending a sponsor. And, according to Val E. Limburg, these fears are not completely unfounded.  To demonstrate his point, Limburg writes about a Western television station's decision to air a series of investigative reports on auto safety:

> One of the segments contained information that some auto makers had not introduced safety features in their new models. They could have, but decided not to on the basis of costs and eventual dividends for their stock brokers.

> One of the local advertisers was a car dealer who learned of the series, at least one segment of which was not flattering to the product he sold. He requested the station's sales representative to kill the series or lose his business. The sales rep respectfully asked the news director to drop the series. The news director refused, indicating that the request amounted to censorship and a second-guessing of his own news judgment.

> The general manager was called in to arbitrate the dispute. Was the station to keep its good client and cut the series, or keep what the news director considered to be its "news integrity" and carry the reports?

> After considerable discussion, the general manager opted to keep the news series. The car dealer withdrew his advertising.[23]

> Thus, one can see that there is often a price to pay for integrity in broadcast news dissemination.

The previous discussions of TV journalism limitations (time, visual aesthetics, emphasis on entertainment, and economic restraints) should not be viewed as an indictment against the electronic news industry nor as an endorsement of Agnew's 1969 speech. Rather, it is intended as a foundation from which the unique world of television journalism can be explored. It can also serve as a basis for analyzing the evolution of TV ethics, today's ethical dilemmas, and the impact of technological and economic developments on future ethical decisions. Accordingly, this book will be divided into four parts:

1. the evolution of various television programming codes and journalistic guidelines;

2. an ethical survey of general standards among today's TV news directors;

3. specific case study problems in TV news ethics; and

4. summary and general conclusions about TV ethics in the 90s and beyond.

## NOTES

1. See Appendix I for the complete text of Agnew's Des Moines, Iowa, speech.

2. Barry G. Cole, ed., "Quality of News," in *Television* (New York: MacMillan and Co., 1970), p. 9.

3. Spiro T. Agnew, "Television News Coverage," in *Vital Speeches of the Day* (December 1, 1969), p. 98.

4. *Ibid.*, p. 99.

5. *Ibid.*, p. 101.

6. Cole, p. 9.

7. Cole, p. 9.

8. Frank Judge, "Critic Questions News Practices," *The Detroit News* (February 18, 1973), p. 67.

9. *Ibid.*

10. Verne Gay, "CBS Staged Vietnam Atrocity, Says Book by Army Historian," *Variety* (July 19, 1989), pp. 1-2.

11. Shirley Biagi, *Media/Impact: An Introduction to Mass Media* (Belmont, CA: Wadsworth Publishing Company, 1988), p. 352.

12. Paul Harris, "Ethics Issues a Hotter Topic than Usual for RTNDA," *Variety* (September 6, 1989), p. 66.

13. Hugh Downs, *On Camera: My 10,000 Hours on Television* (New York: G.P. Putnam's Sons, 1986), p. 194.

14. Barbara Matusow, *The Evening Stars* (New York: Ballantine Books, 1983), p. xiv.

15. Biagi, p. 346.

16. Maurice R. Cullen, Jr., *Mass Media & the First Amendment* (Dubuque, IA: Wm. C. Brown Company Publishers, 1981), p. 341.

17. "Rather Concerned over Afghan Flap," *TV Guide* (January 27, 1990), p. 43.

18. Tamotsu Shibutani, *Improvised News: A Sociological Study of Rumors* (Indianapolis-New York: The Bobbs-Merrill Co., 1966).

19. Jan R. Costello, "Exploiting Grief: Restraint & The Right to Know," *Commonweal* (June 6, 1986), pp. 327-329.

20. James McGregor, "How ABC's Av Westin Decides What to Show on the Evening News," *Wall St. Journal* (November 22, 1972), p. 1.

21. Michael Ignatieff, "Is Nothing Sacred? The Ethics of Television," *Daedalus* (Fall 1985): pp. 57-58.

22. "Media Probes: TV News," produced by WQED-TV (Pittsburgh, PA), 1982.

23. Val E. Limburg, *Mass Media Literacy: An Introduction to Mass Communication* (Dubuque, IA: Kendall/Hunt Publishing Company, 1988), pp. 460-461.

# 2

▼
▼
▼
▼
▼

# The Evolution
# of TV Journalism and
# Broadcast Ethics Codes

During the ten years prior to 1989, network news shares had reportedly dropped from 76 to 59.[1] While this downward trend has also been found in general prime-time television, it is an especially disturbing sign for TV journalism. This is because, for many years, televised news has been the sole source of information for many Americans.

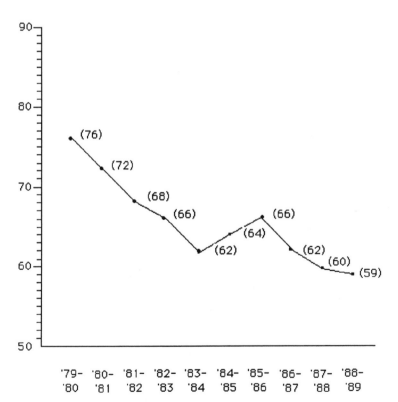

▶ *Figure 1.* *Total News Shares for ABC, CBS, and NBC—1979–1989. Adapted from data supplied by Bob Knight, in "Network Evening News Shares Plummet From 76 to 59 Since 1979–1980,"* Variety *(September 20, 1989), p.124.*

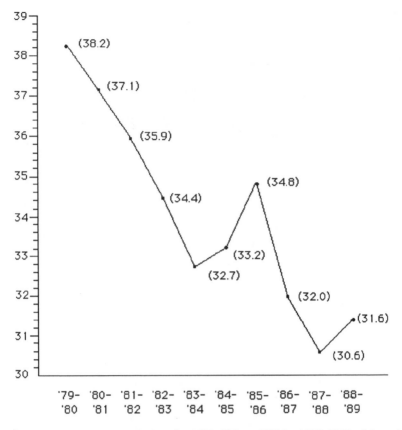

▶ *Figure 2. Total News Ratings for ABC, CBS, and NBC—1979–1989. Adapted from data supplied by Bob Knight, in "Network Evening News Shares Plummet From 76 to 59 Since 1979–1980,"* Variety *(September 20, 1989), p.124.*

Many media critics have argued that there are several explanations for this decline in viewership, most notably:

1. the increase in time devoted to national news by local TV stations;
2. the increase in length of early fringe local news shows, generating a viewer fatigue factor;
3. the powerful advent of cable news as a competitor;
4. the recent interest in tabloid TV as a substitute for hard news; and
5. the possibility that audiences are turned off by what they perceive as biased, slanted reporting.

While all of the above theories seem logical, this book is most concerned with the apparently deteriorating public perception of TV news over the last 20 years. As mentioned previously, electronic journalism possesses some inherent limitations. However, can these limitations be better managed? One way of finding out is to explore the origins of TV news: its purpose, its ethics, and its possibilities.

## THE EARLY DAYS OF TELEVISED JOURNALISM

The first nightly news service was inaugurated by CBS-TV on August 15, 1948.[2] It was a singularly unimpressive affair—no sponsors were even found for the show—but network executives thought that some sort of news presentation would be a good move toward improving CBS's image as a public servant. They also felt that presenting such programming might ease possible future governmental interference with the new medium. In short, TV news in the late 1940s existed as more of a pleasant companion to the business of programming rather than as an integral part of broadcasting. In fairness, however, it must be noted that TV would have been facing an uphill battle for popularity, no matter how committed programming executives were to the news.

In 1948, radio was still seen as the golden medium; most Americans received their information from it as well as from newspapers and movie house newsreels. Part of the reason for the popularity of radio news was its enjoyable style. Audiences loved to listen to the dramatic voices of H.V. Kaltenborn, Boake Carter, and Gabriel Heatter. They also wanted to hear the commentators' personal opinions about the issues of the day. These familiar personalities made listeners feel knowledgeable and in control of world affairs. According to Barbara Matusow, Walter Winchell was perhaps the most noteworthy in this respect; he often attracted over 25 million loyal fans a night with his unique blend of gossip, innuendo, and biased reporting.[3]

Given this context, how could televised news possibly compete with radio's established popularity? Nevertheless, CBS, using Douglas Edwards as its announcer, stepped cautiously into the area of TV journalism, and tried to woo audiences away from the more established news sources of the late 40s. At NBC, broadcast news developed somewhat differently, although along parallel lines. Like CBS, network attitudes were dubious at best about TV journalism.[4] However, in this case, the producers were at least fortunate enough to solicit an advertiser for the show—R.J. Reynolds Company—despite the fact that sponsored news was not always a bonus (as NBC soon discovered). Before signing on, Reynolds reserved the right to approve of the announcer (John Cameron Swayze) as well as the format of the show. Matusow comments:

> Advertisers got a lot for their money in those days. Besides the commercials
> . . . every Swayze show ended up with a long, lingering shot of a lighted
> Camel, smoke curling toward an ashtray. Swayze, always eager to please a
> sponsor, offered to carry a pack of Camels around with him, even though he
> didn't smoke, but the Camel people felt it wouldn't be necessary.[5]

Reynolds even had a "gentleman's agreement" with NBC to avoid broadcasting any pictures of people smoking cigars. Winston Churchill was the only exception to this dictate.

By 1951, CBS and NBC had begun to consider TV newscasting a bit more seriously.[6] Both networks had coaxial cable and microwave relay systems; both released 15 minutes of news each day; and both relied heavily on newsreel companies for the respective film footage. CBS's Douglas Edwards and NBC's John Cameron Swayze each sat at desks in a small studio, reading their wire service copy into bulky ribbon microphones—no one even contemplated the possibility of broadcasting from outside a studio. Still, according to Desmond Smith, some competitive urges between the two networks began to emerge:

> . . . some smart young director decided to hang a clock in the background,
> just so the audience would know it was all happening before their eyes.
> Then an even smarter young director on a rival network added a second
> clock with the time on the West Coast . . . Pretty soon it degenerated into a
> game of "Can you top this?" If Douglas Edwards at CBS had three clocks
> and a calendar, then NBC's John Cameron Swayze, a dapper man who
> usually wore a red carnation in his buttonhole, had to have a map. So maps
> proliferated. Not to be outdone, Edwards' producer added a world globe, a
> black telephone (evidently in case Edwards had a sudden urge to call the
> President), In and Out baskets, and an executive pen set.[7]

Admittedly, once the *show biz* aspect of television news was begun, it would be difficult to reduce, much less to eliminate. However, it is important to recognize that at this time, television was also developing serious journalistic credentials despite sponsor influence, entertainment urges, and technological constraints. Most media historians refer to this period as "the Murrow era;" in truth, it belongs to other people as well. But it cannot be denied that Edward R. Murrow's charismatic vision contributed most significantly to the dramatic rise in credibility of TV journalism from the late 50s to the early 60s.

## THE MURROW ERA

Edward R. Murrow's first telecast for CBS television occurred on November 18, 1951, at 3:30 P.M.[8] However, he was no stranger to most Americans. Few listeners did not recognize the famous voice that had guided them through World War II with reports such as this:

> . . . men with white scarves around their necks instead of collars . . . dull-
> eyed, empty-faced women . . . Most of them carried little cheap cardboard
> suitcases and sometimes bulging paper shopping bags. That was all they
> had left . . . A row of automobiles, with stretchers racked on the roofs like
> skis, standing outside of bombed buildings. A man pinned under wreckage
> where a broken gas main sears his arms and face . . . the courage of the

people; the flash and roar of the guns rolling down streets . . . the stench of air-raid shelters in the poor districts.[9]

Murrow had grown up in a small community on Puget Sound in Washington state. After graduating from high school, he went on to college to become an educator; however, the onslaught of World War II changed his aspirations. He became a reporter instead. And, once deciding upon journalism as his future career, Murrow lost no time in gaining recognition. His rich voice, in addition to his uncanny ability to capture the mood of the moment, was immediately successful. He quickly became a radio broadcast celebrity. In 1937, CBS sent him to London to schedule broadcasts with various European officials. By 1938, listeners sat anxiously by their radio receivers to experience, through Murrow's eyes and ears, a war thousands of miles away.

While Edward R. Murrow's journalistic abilities were legendary, so also were his talents in choosing qualified staff people as reporters. In the late 1930s and early 40s, William S. Paley, president of CBS, commissioned him to hire a cadre of radio journalists for war coverage in Europe. Murrow found some of the best talent in the world, including William L. Shirer, Howard K. Smith, Robert Trout, Larry LeSueur, Eric Severeid, Richard C. Hottelet, David Schoenbrun, and Charles Collingwood. This brilliant team of men actually developed the art of radio broadcasting in Europe, and elevated it to a legitimate form of news reporting in the United States. While NBC was also heavily involved in wartime reporting (many journalists rode the same jeeps and trucks as Murrow's CBS team), no one could compete with Murrow's stature—he . . . and his men . . . had made World War II real for most Americans.

As a result of this dramatic turn of events, radio news gained a measure of credibility not previously known to the medium. According to a poll taken by the National Opinion Research Center at the University of Chicago in 1945, for example, 61% of those surveyed said they received most of their news from radio as compared to 35% who favored newspapers.[10] This certainly was due, in part, to Edward R. Murrow, who epitomized the ideal of journalistic reporting with integrity as well as feeling.

Meanwhile, at CBS, William Paley was already contemplating the move from radio to television. He immediately thought of Murrow and his team as the perfect sources to introduce TV network news after the war. Unfortunately for Paley, however, Murrow was not in agreement with this decision, at least for awhile. David Halberstam described Murrow's dilemma in *The Powers That Be*:

> From the start Murrow regarded television with suspicion. It was there, it was clearly going to be important, and he was a communicator, and whatever else it was, it was clearly a powerful forum for communication. But, like his colleagues in print, he felt that television, after all, was somehow another world, closer to the world of show biz than to the purity of ink. In addition, he was not sure it was a good conduit for the transmission of

ideas, and he was ill at ease with the sheer force of it, with what he suspected was a tendency to overdramatize, and a likely incapacity for dealing in subtleties.[11]

Murrow also felt that he might not be as successful in television as in radio. For one thing, he knew that radio was much more intensely personal than TV; he also was singularly unexcited about working with a producer. Most of his doubts dissipated, however, when Paley agreed to pair Murrow with his longtime radio associate, producer Fred Friendly. Together, they produced several documentary and interview series, including "See It Now" and "Person To Person." By the early 1950s, Murrow was ensconced in TV news, and his presence created a newfound credibility in the medium. Many of his respected radio colleagues soon followed Murrow's lead into television.

"See It Now" established a new standard in aggressive journalism for TV news, which now showed bold innovation in both content and presentation. Originating from a studio control room to add immediacy to the broadcast, Murrow and Friendly addressed topics such as America's neglected migrant workers, freedom of speech in Indianapolis, and the controversial discharge of Airman Milo Raduvolich. However, most memorable of all was Murrow's dogged investigation of Senator Joseph McCarthy and his anticommunist witch-hunt.

Murrow had wanted to feature McCarthy's blacklist on one of his broadcasts in 1951, but CBS had been hesitant to air such a controversial issue. Added to the network's concern was the fact that the broadcast would be aired live (since videotape was not in use at this time)—a poor performance could mean government repercussions as well as viewer outrage. Nevertheless, Murrow and Friendly started research for the show on their own in 1953, bought print ads promoting the broadcast with their own money (sans the CBS logo), and kept the idea of their show a secret for as long as possible.

The content of the program was simple: Murrow and Friendly showed newsreel footage of McCarthy in action, then analyzed his charges and unethical tactics. Murrow let McCarthy speak for himself and then ended the program with this comment:

> We will not walk in fear of one another, we will not be driven by fear into an age of unreason. If we dig deeper in our history and our doctrine and remember that we are not descended from fearful men, not from men who feared to write, to speak, to associate, and to defend causes which were for the moment unpopular . . .[12] This is no time for men who oppose his methods to keep silent. We can deny our heritage and our history but we cannot escape responsibility for the result.[13]

Viewer reaction to the broadcast was overwhelmingly favorable, and Murrow and Friendly were congratulated for their courageous endeavor. Most significantly, it was apparent that political commentary was as welcome on television as it had been on radio.

## TV JOURNALISM IN THE 1950s, 1960s, AND 1970s

Although Edward R. Murrow and his team at CBS attracted a large number of television viewers, NBC was not standing by idly. In the mid-50s, Robert Kintner had become president of the network, and one of his primary missions was to make NBC news top in the ratings. Kintner was a journalist by trade; a broadcast executive by circumstance; and, by all accounts, a hard-driving, aggressive man by nature. He pledged that NBC would never be beaten on a news story again. He also teamed up a soon-to-be legendary duo of newscasters, Chet Huntley and David Brinkley, to co-anchor the evening broadcast in 1956.

By the end of the decade, both NBC and CBS had grown so large as to include about 200 affiliates each.[14] This meant that news could be covered instantaneously throughout the United States. It also meant that TV news had begun to play an integral role in education, politics, and modern history. The latter aspect began to trouble critics of mass communication.

On May 9, 1961, newly appointed FCC Chairman Newton Minow voiced his concerns about broadcasters' responsibilities to the public:

> Your license lets you use the public's airwaves as trustees for 180 million Americans. The public is your beneficiary. If you want to stay on as trustees, you must deliver a decent return to the public—not only to your stockholders . . . Ours has been called the jet age, the atomic age, the space age. It is, also, I submit, the television age. And just as history will decide whether the leaders of today's world employed the atom to destroy the world or rebuild it for mankind's benefit, so will history decide whether today's broadcasters employed their powerful voice to enrich the people or debase them.[15]

Minow's caution not withstanding, televised news had catapulted itself into a position of power, prestige, and high credibility by the early 60s. Consider, for example, that John F. Kennedy's 1960 presidential victory was due in part to his televised debates with Richard Nixon. There is no question that based on this early success, Kennedy later used television as a forum to conduct live press conferences and speak to his national constituency.

As for other developments, in July 1962, Americans witnessed the launch of a new communications satellite, *Telstar I*, and for the first time in history, worldwide news as an immediate commodity became possible. This technological phenomenon became all too apparent when Kennedy used satellite television to deliver an ultimatum to Fidel Castro during the Cuban missile crisis. It also played a major part in delivering the horrific news of Kennedy's assassination in 1963.

In addition to telecasting the trauma of assassination, 1963 was also designated as a significant year for other TV news developments: ABC seriously entered the world of broadcast journalism; network programmers extended the 15-minute evening newscast to one half hour; and Walter Cronkite replaced Douglas

Edwards as CBS-TV's primary anchorman. By 1965, all three networks were telecasting in color and commercial satellites filled the universe. And by the late 60s, reporters with newly developed electronic news gathering technology covered stories from Vietnam to Atlanta, Georgia, from Tokyo to Washington, D.C., and from Newark, New Jersey to the Apollo spacecraft on the moon. In addition, the most intensive protests in American history appeared between 1968 and 1974, as anti-Vietnam war marchers and civil rights leaders dominated the TV screen. From all regions of the country, audiences viewed college student demonstrations, political convention riots, police brutality, prison unrest, and political assassinations. No one who lived through these turbulent times could ever forget the image of violence at the 1968 Chicago Democratic Convention or the excruciating film footage of the My Lai massacre. In fact, television reporting during this time affected viewers like news reports never had before.

Most media critics would agree that the conflict in Vietnam played a pivotal role in most of the unrest shown on TV during the late 60s and early 70s. Billed as "the first televised war" in history,[16] Americans often consumed network news combat footage along with their evening meals. Also, news anchors at NBC, CBS, and ABC could not help but editorialize after presenting graphic footage of

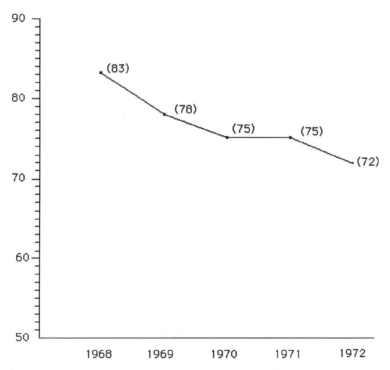

▶ *Figure 3. Total News Shares for ABC, CBS and NBC—1968–1972. Adapted from data given in* Television/Radio Age *(October 30, 1972), p. 29.*

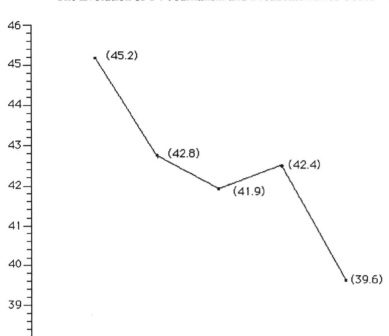

▶ *Figure 4. Total News Ratings for ABC, CBS and NBC—1968–1972. Adapted from data given in* Television/Radio Age *(October 30, 1972), p. 29.*

napalm bomb damage, frantic Vietnamese women and children, and demoralized U.S. troops. In fact, some experts contend that Walter Cronkite played a significant role in altering Lyndon Johnson's defense policy by his negative commentary after the disastrous results of the Tet offensive.[17] While Cronkite was probably only one voice among many, it still appears that government officials resented the freewheeling activities of broadcast journalists in the Vietnamese war zone. Since that time, the U.S. Defense Department has forbidden press entry into Grenada, Panama, and other places of combat, claiming that unauthorized video footage might affect the outcome of the military mission.

TV reporting also affected American politicians as it never had before, and many of these politicians sought measures to control its tremendously growing power. Most vocal among TV's detractors was Spiro T. Agnew, whose 1969 Iowa speech (discussed in Chapter 1) invited viewers to question TV news techniques. Later, then President Nixon formalized Agnew's criticisms by creating a watchdog Office of Telecommunications Policy, under Clay T. Whitehead. The OTP's main mission was to extend more informal censorship on television programming, especially in the area of news.

It must be stated at this point, however, that television news was not completely innocent of the charges raised against it. By the mid-70s, stories abounded about such questionable activities as simulated drug busts, staged demonstrations, and checkbook journalism. Concerned critics and journalists realized that such practices would endanger TV news credibility, and hence, its existence. In fact, the overall viewing of network news showed a significant decline from 1968 to 1972, in part due to a growing lack of credibility in electronic journalism.

To adequately overcome these inadequacies, broadcast journalists increased their attempts to improve their product quality. Unfortunately, however, the realities of time constraints, technological developments, and economic concerns still prevailed in both network and local news.

## TV JOURNALISM IN THE 1980s

Despite constant accusations of slanted reporting, biased coverage, and rampant sensationalism, TV news continued to be a powerful force in the early 1980s. At the start of the decade, each network had apportioned at least $200 million a year for its news department,[18] and unequivocally, broadcast journalism continued to set the agenda for political campaigns, world issues, economic priorities, and environmental affairs. In addition, the news anchor[19] continued to rise as a hallowed spokesperson for the American public—a phenomenon that fostered the notion of news as *show biz* rather than news as a public service. News anchors began to receive multi-million dollar salaries and multi-year contracts, which had been negotiated by high-priced talent agents. Anchors were also given greater editorial privileges as well as the right to hire and fire producers at whim. In short, both networks and local stations viewed the television news anchor as a representative for the public image of that station/network. More specifically, viewers were seen as more likely to tune into a newscast because they liked Rather, Brokaw, or Jennings, not because of a technological innovation such as satellite feeds, digital graphics, stereo TV, or high-speed Beta tape. Further, according to audience researchers, persons who began their evenings with a favorite news personality often extended their viewing loyalty to the highly profitable prime-time period of the same network.

Thus, while all networks (and most major market stations) used similar state-of-the-art technology in their broadcasts, the measurement of their success or failure often seemed to lie in the personality of the network/station rather than in its technological superiority or electronic news gathering techniques.

It is important to note at this point, however, that the need for personality oriented news was not created solely for prime-time profitability. In fact, in the early 1980s, many broadcast executives saw the move as one of sheer survival, given the changing nature of the television marketplace and its competitors. By 1981, networks had watched their previous 95 share (in the late 1950s and early 1960s) diminish to an 84; research analysts were predicting a continued drop throughout the decade (which, indeed, happened).[20] The major reason for audience shrinkage was attributed to the new technology and its accompanying rise in fragmentation. In other words, viewers were no longer dependent solely on the networks or their affiliates for television news and information. Other alternatives abounded, including CNN, FNN, ESPN, and The Weather Channel, as well as increasing news programs on independent stations.

Unfortunately, the continued pressure of dwindling audiences caused the business of news gathering to become more "business" and less news gathering. Viewers seem to be more fickle than ever, flocking to personality-oriented, tabloid news magazines[21] like "A Current Affair," "Inside Edition," and "Hard Copy," as well as the pseudo news programs such as "Rescue 911" and "Unsolved Mysteries." As a result, local and national news directors have felt increased pressure to behave in similar fashion, creating more difficult ethical dilemmas than have ever been felt in the world of televised journalism.

## TV NEWS AND ETHICS

The evolution of coded ethics in TV journalism has run a course very similar to that of the broadcast programming history discussed earlier in this chapter. Ethical codes in information presentation had first been developed during the early years of radio. They were later simply adapted to include television. Today, most TV and radio stations still consult the Radio-Television News Directors Association and Society of Professional Journalists—Sigma Delta Chi (SPJ-SDX) for ethical guidelines in presenting the news. (The RTNDA and Sigma Delta Chi codes of broadcast news ethics are found in Appendixes B and C at the end of this text.) However, prior to 1982, the National Association of Broadcasters also played a significant role in ethical news broadcasting. The subsequent demise of the NAB Code was due in part to increased technological developments, governmental influence, and the threat of economic reprisal from advertisers—a situation which seems to parallel the history of TV news programming. Since 1982, the NAB has operated without a code, and at this time, there seems little likelihood that a code will be reinstituted.

The National Association of Broadcasters was founded in 1922 as a nonprofit organization to:

> . . . foster and promote the development of the arts and aural and visual broadcasting in all its forms; to protect its members in every lawful and proper manner from injustices and unjust exactions; to do all things necessary and proper to encourage and promote customs and practices that will strengthen and maintain the broadcasting industry to the end that it may best serve the public.[22]

Another of the NAB's first major missions was to diffuse the hostile relationship between newspaper editors and broadcast journalists, which had begun in the early 1920s and had escalated to epic proportions by 1930.

It happened innocently enough. Radio news had come into being very uneventfully in the mid-20s, mainly because it was one form of live broadcasting that didn't require loads of actors and actresses, technicians, or sound effects people. Further, by presenting daily news briefs, radio executives claimed that they were providing a needed service to their audiences, thus fulfilling the Federal Radio Commission's request that they address the "public interest, convenience, and necessity" in all programming. Surprisingly to everyone, the newscasts became very popular, and, by 1930, regularly scheduled news shows and radio commentaries made up a significant portion of the broadcast day, both at network and local levels.[23]

Unfortunately, however, this new boon to information was not welcomed by everyone. Most notable among radio's detractors were many newspaper owners and publishers, who saw this latest technological development as a threat to their own existence. The newspaper industry, it seems, had witnessed shrinking ad revenues for several years, due to radio's popularity. Understandably, it was not anxious to see this trend continue.

Thus, in what was later labeled "the press-radio war of 1933," many news publishers refused to print broadcast schedules, and convinced wire services such as AP, INS, and UP to limit access to radio news reporters. Further, they threatened to curtail or eliminate news coverage for any sponsor of any program on CBS, the recognized leader of radio news at that time.[24] These attempts at sabotage greatly hampered network news presentation, and ultimately forced radio to establish its own sources of information.

In the midst of "the press-radio war," the National Association of Broadcasters was asked to mediate the conflict between networks and publishers. By the end of 1933, NBC, CBS, and the NAB signed a statement, agreeing to:

1. limit the amount of news produced, airing no more than two five-minute broadcasts each day;[25]
2. limit the number of words for each individual news item (no more than thirty);[26]
3. refrain from editorializing on any news item less than twelve hours old;[27]
4. encourage audiences to read more newspapers; and
5. to create a Press-Radio Bureau, which would serve as a watchdog to guarantee all previous agreements stated above.[28]

The compromise endorsed by the NAB, the networks, and newspapers (sometimes referred to as "the Biltmore agreement," for the hotel where it was signed) broke down within months, however, for several reasons:

1. the Press-Radio Bureau soon became ineffective, due to the growing number of competing radio wire services which began to emerge;
2. the Biltmore agreement did not include all radio station owners (since many were not network-affiliated and were not represented at the meeting)—they did not feel bound by the compromise and presented as much news as they could, using as many sources as possible; and

3. the advent of overseas news about impending war in Europe was being presented faster and better on radio than in the newspapers.

Thus, by the late 1930s, radio had finally been freed from the yoke of the newspaper industry. Station and network owners subsequently began to tighten their own ranks, and the National Association of Broadcasters grew, making it much stronger than it had been before. The power of the NAB continued to increase in 1951, when television station owners became members of the organization as well. During this time, the NAB's voluntary code of programming standards became legendary—no respectable radio or TV station wished to be dropped from membership because of code violations—and the model of broadcast ethics seemed to be firmly intact.

The growing status of the NAB never became more apparent than in the 1960s and early 1970s, when the Association's annual conventions were often keynoted by such luminaries as Newton Minow, Senator Mike Mansfield, and Dean Burch. Most memorable, however, was the historic confrontation between then President Richard Nixon and CBS's Dan Rather at the conference in Houston, March 1974. The Watergate scandal had already begun to unravel, and Nixon and his advisors felt that one successful way to stem the rising tide of public mistrust was to gather sympathy through an informal political stage. According to David Halberstam, the annual meeting of the NAB:

> . . . had all the necessary ingredients. It was in the South, perfect for their purpose since the South was the citadel of conservatism and it would be easier to get enthusiastic crowd responses there than elsewhere, a chance to get scenes of the President being cheered for the evening news shows. The setting was even better because in the public's mind "broadcasters" were authentic Nixon antagonists, were men like Cronkite or Rather or Schorr or Carl Stern or Fred Graham, enemies of Richard Nixon, men who had covered all of Watergate and caused all these problems.[29]

Dan Rather, among others, was dubious about Nixon's willingness to appear at the convention. In fact, he had promised himself he would not "rise to the bait" by asking any questions of the President. Rather nearly succeeded in his vow . . . until Nixon commented that he had fully cooperated with Watergate Special Prosecutor Leon Jaworski. Halberstam continues:

> . . . reluctantly, Rather rose, and as he did, and as he tried to identify himself, he seemed to hear an enormous amount of booing . . . the arrival of the moment of truth, here was the confrontation that everyone wanted, the last inning of a close baseball game, the last two minutes of a football game, the ultimate confrontation between bull and bullfighter . . . Richard Nixon, hearing the noise, warmed to it—"Are you running for something?" he asked . . . He [Rather] answered the President—"No sir, Mr. President, are you?"[30]

Quite coincidentally, then, the 1974 National Association of Broadcasters Convention exposed TV and radio newscasters to the public more than they'd ever been exposed as a group before. It also seemed to exploit them, inadvertently presenting national reporters as vicious expatriates with power-hungry motives. Never before or since has the NAB been spotlighted more brightly; never before or since has its credibility been questioned as intently, either. In March 1974, Nixon had succeeded in hurting his perceived nemesis. Broadcast journalists ultimately won the war, however, by exposing the Watergate scandal completely and causing Nixon's resignation from office.

In 1982, for quite different reasons, the NAB Code was abolished after a dissatisfied advertiser convinced the U.S. Justice Department to bring anti-trust action against the Association's suggested limitation on advertising time.[31] As yet, the Code has not been re-instated nor are there any plans to do so. Regarding other ethics codes in other associations, both RTNDA and SPJ-SDX modified their provisions in 1987 to exclude censuring journalists for unethical practices, among other things. But as Pasqual et al. suggest in *Mass Media in the Information Age*:

> Ethical codes are not laws; they are guidelines lacking enforcement power. Ethical codes thus suggest rather than command. When a society feels strongly enough about an ethical issue, it may enact a law. That does not mean that what is legal is ethical. The ethical and legal considerations may be the same or different. Usually there are more ethical perspectives than legal ones on any given issue. Certain topical areas likewise attract more ethics debates than others.[32]

As will be shown in the next two chapters, ethical beliefs in broadcasting are usually individualized and personal, despite general policy agreements. Furthermore, TV news directors may often encounter stumbling blocks which can affect the presentation of information despite their high ethical standards. Finally, as Pasqua and his colleagues have noted above, certain news items inherently raise more ethical questions than others.

### Notes

1.  Bob Knight, "Network Evening News Shares Plummet From 76 to 59 Since 1979-80," *Variety* (September 20, 1989), p. 124.

2.  This chapter will focus primarily on network news, since most available historical information deals with national broadcasts versus local TV journalism. The major reason for the profundity of national news information is simple: local news did not become a serious consideration in TV broadcasting until at least a decade after network news began. Also, it should be noted that local news follows much the same pattern as network news. Thus, network news history serves as an excellent blueprint to study for both local and national news broadcasting.

3.  Matusow, p. 51.

4.  Reuven Frank, president of NBC News from 1968 to 1973 and again from 1982 to 1984, most closely describes the feeling of the time: "I asked this guy how

come he hired me. 'Well,' he said, 'Nobody in radio who is worth anything thinks it's gonna last.'" (Desmond Smith, "TV News Did Not Just Happen—It Had to Invent Itself," *Smithsonian* [June 1984], p. 76).

5 . Matusow, p. 65.

6. ABC did not really enter into the arena of network news until 1953. Even after its entry, however, it was not viewed as very serious competition to NBC and CBS.

7. Smith, pp. 78, 80.

8. Biagi, p. 121.

9. A.M. Spencer, *Murrow: His Life and Times* (New York: Freundlich, 1986), p. 168.

10. Matusow, p. 54.

11. David Halberstam, *The Powers That Be* (New York: Dell Publishing Co., Inc., 1979), p. 192.

12. *Ibid.*, p. 205.

13. Smith, p. 87.

14. *Ibid.*, p. 88.

15. Newton Minow, *Equal Time: The Private Broadcaster and the Public Interest* (New York: Atheneum, 1964), p. 51.

16. "Television," produced by WNET-TV (New York) and KCET-TV (Los Angeles) in association with Granada Television (London), 1988.

17. *Ibid.*

18. Matusow, p. 304.

19. The term *anchor* actually was coined during the 1952 national political conventions, when Sig Mickelson and Don Hewitt at CBS wanted a strong presence in the reporting booth. According to Barbara Matusow, "Hewitt compared the arrangement to a relay team, where the strongest runner, who runs the final leg of the race, is called the *anchorman*. It was not a particularly elegant expression as applied to television, but it stuck" (p. 74).

20. Matusow, pp. 306-307.

21. Jack Dempsey, "More Mags Will Fly in the Fall; Too Much of a Bad Thing?" *Variety* (April 12, 1989), pp. 79, 98.

22. Craig T. and Peter G. Norback, eds. *TV Guide Almanac* (New York: Ballantine Books, 1980), p. 483.

23. Thomas M. Pasqua, Jr., James K. Buckalew, Robert E. Rayfield, and James W. Tankard, Jr., *Mass Media in the Information Age* (Englewood Cliffs, NJ: Prentice-Hall, 1990), p. 33.

24. *Ibid.*

25. Matusow, p. 51.

26. *Ibid.*

27. *Ibid.*

28. Pasqua et al., p. 34.

29. Halberstam, p. 974.

30. *Ibid.*, pp. 975-976.

31. Limburg, p. 458.

32. Pasqua et al., pp. 263-264.

# 3

# Ethics and Today's Television News Director

When exploring the concept of ethical behavior, most academic scholars feel it is important to create a structure from which to observe the ideal. Media critics have assumed several different approaches to the topic, but for purposes of this text, one of the most valuable tools of analysis comes from Dr. Ralph Potter of the Harvard Divinity School.[1] Potter includes four dimensions of ethical analysis in his work: definition, values, principles, and loyalties. He has also arranged these dimensions into a formula of progression, aptly entitled *Potter's Box*. By using Potter's definitions as well as his diagram, it is possible to discover patterns of decision-making in every day life. This is especially valuable when looking at TV news directors and the problems they encounter when producing electronic newscasts.

Potter begins by *defining* the situation, according to the importance of the event and the options available for covering it. This allows news directors to see the event as objectively as possible, without needless encumbrances.

Next, according to Potter, news directors must determine their own *values* in the matter. As Christians, Rotzoll, and Fackler state in *Media Ethics:*

> Any single decision involves a host of values and they must be sorted
> out. . . . We may judge something according to
>    aesthetic values (harmonious, pleasing),
>    professional values (innovative, prompt),
>    logical values (consistent, competent),
>    sociocultural values (thrift, hard work), and
>    moral values (honesty, nonviolence.)[2]

Each value determines the manner in which the news director will approach the story.

This value-determination process is most necessary before the news executive can move on to consider the *principles* of the station policy (broadcasting the truth, respecting an individual's rights, selling ad time to make money, etc.), which may conflict in some ways with the previously stated values.

Finally, after news directors weigh their own values with the station's goals and objectives, they can then make some decisions about *loyalties*, whether it be to themselves, to the individuals in the news story, or to the station policy (if each is in conflict with the other).

Potter's framework of ethical dimensions is most useful when demonstrating a "relationist" perspective to broadcast journalism (see Introduction), and it will be used more specifically in the case studies found in the next chapter. However, it is important to introduce Potter's Box in general terms as well as in individual cases; it provides a strong structure for ethical reasoning.

After reviewing the history of television journalism found in Chapter 2, it becomes evident that today's news directors may not always be able to make easy decisions about which stories are most appropriately suited to each day's broadcast. Further, they may have difficulty positioning themselves between their own personal ethics and the unique challenge of running the business of TV over public airwaves.

The purpose of this chapter is to explore the basic ethical standards of news directors in the 1990s (using Potter's Box), because their beliefs, attitudes, and values about television journalism are most significant. Also, these insights might allow viewers to better understand the process of choosing daily news content and story placement in local as well as network news programs.

To investigate today's ethical news standards in television, the author sent detailed surveys to 240 news directors across the United States. The questionnaires were mailed in November 1989, and followed the distributional pattern indicated below:[3]

1. eighty executives were from large ADI markets,[4] eighty from medium-sized ADI markets,[5] and eighty from small ADI markets;[6]
2. network O & O stations, network affiliates, and independent television stations were all represented; and
3. seven different geographic regions were covered.[7]

The survey posed several questions pertaining to general ethical standards as well as judgments about certain TV news stories in 1989. On the facing page is the list of questions. To compare your responses with those given in the survey, spaces have been provided for notes.

Only 9.17% of the selected TV news directors returned completed responses. However, these responses were generally quite detailed and certainly diverse. Moreover, the people who took time to answer the questionnaires were earnest, articulate, direct, and most helpful. Thus, it can be said that this body of responses, if not authoritative, is generally indicative of the priorities, perceptions, and concerns of TV journalists, and will enable other journalists and scholars to examine their own ethical methods.[8]

The news directors' specific responses to the survey, along with the overall survey findings, are summarized below.

## Ethics in Electronic Journalism
General Questionnaire

How do you define "ethics" (in relation to human rights)?

Speaking as a TV executive, does your definition of "ethics in broadcasting" differ from your overall definition of ethics (stated above)? Why or why not?

In general, which of the following factors are most important when deciding how to present a newscast? (Please rank order them accordingly: 1=most significant . . . 10=least significant.)

_____ national/international security constraints
_____ withdrawal of advertiser support
_____ economic constraints (e.g., cost of production, etc.)
_____ technological constraints (e.g., availability of equipment, satellites, etc.)
_____ First Amendment rights
_____ viewer mail
_____ accuracy
_____ immediacy
_____ viewing aesthetics
_____ time constraints

In your opinion, which news event in the past year has left itself open to the most serious ethical questions? Why?

In your opinion, which news event in the past year has been handled most ethically? Why?

## 1. How Do You Define "Ethics" (in Relation to Human Rights)?

The purpose of this question was twofold: (1) to see if news directors across the country could agree upon a general definition of ethics; and (2) to determine whether most TV journalists saw ethical standards from an absolutist, relationist, or relativist point of view.

Interestingly, everyone agreed that ethical codes were very personal and individualized. They also indicated that qualities such as "morality," "respect," and "fairness" should be a part of every person's moral behavior. However, there seemed to be great differences of opinion when the question implied *universal* values or those that were situationally-oriented.

The majority of respondents (by a 3:1 ratio) felt that ethics were individually derived, but that the basic choices of conduct were universal, permanent, and unchangeable. For example, Steve Snyder (KDLT-TV, Mitchell, SD) was uneasy about the implicit suggestion that personal ethics might be situational:

> There is no definition of ethics "in relation to Human Rights." That is to suggest varying degrees of ethics to fit individual situations. Ethics are an individual's personal code of standards for conduct and cannot change depending on subject matters.

Donna Skattum (WAYK-TV, Orlando, FL) stressed the importance of recognizing the individual's right to define his or her own ethical code. However, she also suggested that there should be a commonality of purpose which remains stable throughout:

> I think ethics basically are human values, and, because of human rights, may vary from person to person. While there are certain intrinsic values that shouldn't vary from person to person, for example, obeying the laws of the land and not killing your neighbor, I strongly feel that individuals have the right to decide the morality of certain issues themselves, and establish their own personal code of ethics.

John McKean (KGGM-TV, Albuquerque, NM) gave the most general definition of ethics. He, too, emphasized personal value-structuring in his response. He also implied that standards, once established, are not subject to change.

> Ethics is the creation and protection of a personal standard of conduct, based on a concept of respect for the dignity and fair treatment of other people, and the maintenance of that standard even when it is not in a person's individual self-interest.

Despite the strong responses by most news directors that ethical standards were permanent and unvaried, some people were less committed to ethical universality. These executives suggested that while ethical codes were extremely important, they were changeable, dependent upon current mores and standards. For example, David Russell (WHKE, Kenosha, WI) explained ethics in this way:

> Ethics is the quantitative measure of respect given to the rights of others.
> Most people operate with an ethical standard somewhere [between] two
> extremes. Some have the ethics of a Mother Theresa—having compassion
> for others, striving to serve and help. Others have the ethics of a sadistic
> rapist-murderer—having absolutely no regard for others or their welfare.
> This standard is uniquely determined either by adherence to a set of prede-
> termined mores or by allowing the situation and current social thought to
> set the standard.

Alan Griggs (WSMV-TV, Nashville, TN) discussed societal laws and the ethical duties one has to respect them:

> The responsibility to treat each other with respect, to adhere in general to
> society's laws, and to maintain, on a daily basis, that the responsibility you
> take for your actions is based on truthfulness and sincerity.

In his statement, however, Griggs seemed to imply that if societal laws should change, so might specific ethical codes.

Liz Talbot (WVTV, Milwaukee, WI) was most situational in her approach to ethics:

> Ethics is the moral responsibility to cover human rights issues (or treat
> human rights) with fairness, virtue, and modern day values.

Talbot, however, emphasized the need for consistency in maintaining some type of ethical code. This comment, incidentally, reflected the opinion voiced by all of her colleagues.

### 2. Speaking as a TV Executive, Does Your Definition of "Ethics in Broadcasting" Differ From Your Overall Definition of Ethics (Stated Above)? Why or Why Not?

The persons responding to this question were almost evenly divided be- tween those who keep their own individual standards while working on TV news and those who change their personal codes to fit the unique business of broadcasting.

Those who responded "no" to the question saw themselves as persons first; broadcast executives second. Skip Haley (WBRZ-TV, Baton Rouge, LA) stated, "How could it be different? 'Right' is not situational in nature."

George Wilson (WOFL-TV, Lake Mary, FL) agreed with Haley. He even questioned the significance of the inquiry, by unequivocally stating that, ". . . competition is not as important as informing the public."

Henry Mendoza (KBAK-TV, Bakersfield, CA) added a bit of professional ethics to his statement, but only insofar as the fact that mass communication organizations allowed broadcasters to see more things than the average person:

. . . I view our role as broadcast journalists as simply being and doing what individuals could do if they could be more than one place at any time. We should serve as the eyes and ears of the public in places and at events they would or should see.

Kent Taylor (KHAS-TV, Hastings, NE) took a very personal perspective on being a news professional. About ethical standards, he stated:

I feel I cannot allow them to differ. We (news broadcasters) must remember that every newsmaker is someone's son, daughter, husband, wife, sister or brother . . . and they could have been us. How would we want to be treated?

Dick Nelson (KCTV, Kansas City, MO) held the same belief, but from a different point of view:

No, it doesn't vary, except that as a journalist one must consider—above everything else—what is "fair" to the public at large. Individual rights may take second place to the right of society to be informed.

Those news directors who responded "yes" to this question clearly felt that personal ethics standards had to change when a newscaster worked for a TV station. However, as the statements below indicate, a "yes" to the question did not necessarily connote a lessening in ethical responsibilities. For example, Brian Moore (WVGA-TV, Valdosta, GA) spoke of his professional duty to consider all sides of a news item before deciding to air it:

My ethics change when I am on the job, because I have to be more stringent and look at each issue more than once and apply my general ethical beliefs to everything, whereas I may not have that "luxury" in everyday life. It is very important that I not do anything to harm anyone on the air because it can have detrimental effects on that individual, my station, and my career.

Alan Griggs (WSMV-TV, Nashville, TN) voiced the same concern . . . and frustration:

. . . there are numerous specific instances when legitimate questions would be raised about modifying the above definition. That's what I'm struggling with as a news director. There are cases when the ethical viewpoint I stated would be clearly instituted and effective. There are many, many other times when it would be unnecessary or, frankly, damaging to adhere strictly to it. Each circumstance or situation should be handled on its own [merit]. It has to be that way. There should be a general code of ethics for each newsroom. Beyond that, never say "never." For instance, how far do journalists go in allowing interested parties to exert influence over us or our stories? Free meals? Those should be specifically forbidden. What if the subject of your story is a personal friend—someone you have grown to like? How do you separate conflicting feelings? Can you? The list can go on and on.

Donna Skattum (WAYK-TV, Orlando, FL) described her professional standards quite clearly. She also stated her news director's responsibility to those reporters who did not follow station policy:

> As a broadcaster, I have always had a certain creed, one that I have required my employees to follow. While their personal morals and ethical code are their own business, when they work for me, their conduct should reflect the station. My ethical code differs from many news directors. If an employee of mine ever shoved a microphone under the nose of a bereaving widow or child after a tragedy, they would be fired. There are boundaries we are not meant to cross, and, while many of my colleagues are guilty of overstepping those borders on a daily basis, no one who works for me will a second time.

John McKean (KGGM-TV, Albuquerque, NM) recounted his feelings about personal versus professional ethical codes accordingly:

> Ethics in broadcasting is the definition and defense of an industry standard of conduct. The standard must be sufficiently flexible to support and nourish a range of individual standards. Ethics in broadcasting must not become associated with rigidity and resistance to change. Principles of fairness apply equally to many forms of communication.

### 3. In General, Which of the Following Factors Are Most Important When Deciding How to Present a Newscast? (Please Rank Order Them Accordingly: 1=Most Significant . . . 10=Least Significant.)

_____ national/international security constraints
_____ withdrawal of advertiser support
_____ economic constraints (e.g., cost of production, etc.)
_____ technological constraints (e.g., availability of equipment, satellites, etc.)
_____ First Amendment rights
_____ viewer mail
_____ accuracy
_____ immediacy
_____ viewing aesthetics
_____ time constraints

While there seemed to be a wide variation among individual responses to this question, the collective rank order of factors to consider before each newscast was surprisingly clear.[8] By far, "accuracy" was seen as the most important characteristic of television journalism; "immediacy" was noted as second most important, with "First Amendment rights" following relatively close behind in third place. "Time," "economic," and "technological" constraints were listed in the middle portion of the ranking, along with "viewing aesthetics." However, "national/international security constraints," "viewer mail," and "withdrawal of advertiser support" were noticeably low on the factor importance scale. This ranking seemed consistent with other, similar works written on the same topic.[9] For specific statistics on the actual survey taken in 1989, please consult Table 1.

### 4.  In Your Opinion, Which News Event in the Past Year Has Left Itself Open to the Most Serious Ethical Questions? Why?

Topics that were mentioned most often in this part of the survey included: simulated news events (like ABC's staged footage of Felix Bloch); news re-creations (like NBC's "Yesterday, Today, and Tomorrow," and CBS's "Saturday Night with Connie Chung"); the Tianamen Square massacre; unauthorized

### Table 1

**Factors Considered in TV News Presentation**
**(1= Most Important; 10= Least Important)***

| | | |
|---|---|---|
| 1 | Accuracy | (1.64) |
| 2 | Immediacy | (2.93) |
| 3 | First Amendment Rights | (3.64) |
| 4 | Time Constraints | (4.46) |
| 5 | Technological Constraints | (5.46) |
| 6 | Economic Constraints | (5.77) |
| 7 | Viewing Aesthetics | (6.31) |
| 8 | National/International Security Constraints | (7.00) |
| 9 | Viewer Mail | (7.83) |
| 10 | Withdrawal of Advertiser Support | (9.58) |

*The figures on the right denote the average of those respondents who ranked these factors in order of importance

video footage of the Sioux City plane crash; the USS Iowa explosion; numerous personal tragedies; the abortion controversy; and various "investigative reports" on political figures. More specifically, respondents had these further comments.

On the abortion controversy, Donna Skattum (WAYK-TV) stated:

> No matter how balanced you try and present the issue, both sides are going to find fault with your coverage. It's also a volatile issue in which few people (including broadcasters) fail to have an opinion, meaning you run a great risk of slanting coverage in one way or another.

On re-creations in news broadcasts, John McLean (KGGM-TV) voiced this opinion:

> Industry has lagged behind advancing technology and has placed visual credibility at risk.

On the Beijing student revolt, Liz Talbot (WVTV) was most concerned about future repercussions for innocent people:

> National media were blacked out from coverage—and when they did get interviews, some [Chinese students] were killed for that coverage. Now Chinese students in America are facing political controversy on when—and/or whether they can go back.

On political investigative reports, Donna Skattum (WAYK) was most articulate, definitive, and possibly prophetic:

> Allegations were enough to cost John Tower the Secretary of Defense position—unsubstantiated allegations . . . a reporter's practice of following a presidential candidate could have conceivably cost Gary Hart the presidency. I'm not saying that the behavior of either man, or any others who fit in this category, is beyond reproach. What I'm saying is that, sometimes at least, we've gone too far. I think it's very difficult to determine where the public's right to know is outweighed by a person's right to privacy. And, I think there will come a time when the tables will be turned on the journalists doing the investigating, when they themselves will become the focus of probes into their personal lives. It will be interesting to see if they're above reproach, and what the public reaction will be, if any.

## 5. In Your Opinion, Which News Event in the Past Year Has Been Handled Most Ethically? Why?

Interestingly, most broadcast executives agreed strongly about which news events in 1989 were presented most ethically. Coverage of Hurricane Hugo topped everyone's list. Brian Moore (WVGA-TV) typified most of the news directors' praise by noting:

> Not only did the newscasters refrain from turning [it] into a sensational story, [they] instead focused on the help that was needed for the area.

Steve Snyder (KDLT-TV) added another dimension to the story's success, by highlighting the advantageous time factor in this case:

> . . . there was enough warning that each network was able to sufficiently prepare and alleviate paranoia over being scooped.

Other news events mentioned in this context were: the sensitive handling of Kitty Dukakis' illness, the opening of the Berlin Wall, and the televangelism scandals. Of the latter, Henry Mendoza (KBAK-TV) wrote:

> I think the media generally showed remarkable restraint in covering the off-the-pulpit activities of the television evangelists. Judging from the information that finally emerged, a lot more aggressive reporting could have been done much sooner in the uncovering of digressions by Bakker and Swaggart.

The results of this survey indicate that TV executives seem to be in general agreement about the need for ethics in the presentation of news events. They also acknowledge possible weaknesses in the broadcast system, and indicate that while general policy-making in ethics is good, it should be flexible enough to allow for specific exceptions to the rule.

The next chapter will address four such special cases—news coverage of a personal tragedy, international news coverage, news anchors as celebrities, and simulation devices in TV journalism. Each case will be analyzed, using Potter's Box as well as some insightful comments from industry professionals.

### Notes

1. The Potter's Box framework was actually developed by Dr. Karen Lebacqz at the Pacific School of Religion. Lebacqz adapted the structure based on Potter's doctoral dissertation, as well as from his essay, "The Logic of Moral Argument," in *Toward a Discipline of Social Ethics,* ed. Paul Deats (Boston: Boston University Press, 1972), pp. 93-114.

2. Clifford G. Christians, Kim B. Rotzoll, and Mark Fackler, *Media Ethics: Cases and Moral Reasoning, 2nd ed.* (New York: Longman, 1987), p. 2.

3. This random selection process was derived from the *Broadcasting/Cable Yearbook '89* (Washington, D.C.: Broadcasting Publications, Inc., 1989).

4. Large ADI markets are defined as those areas with 1 million or more persons.

5. Medium-sized ADI markets are defined as those areas with 500,000 to 1 million inhabitants.

6. Small ADI markets are defined as those areas with less than 500,000 persons.

7. These geographic regions were divided as follows: New England, Middle Atlantic, East Central, West Central, Southeast, Southwest, and Remaining Pacific.

8. In fact, several news directors who returned incomplete questionnaires, actually accompanied them with personal notes. They indicated that there was a station policy to respond only to surveys conducted by professional research firms. The author was appreciative of their honesty and professionalism.

9. Some people responded to all of the factors presented; others responded to the top three, five, or seven only. This deviance explains in part some discrepancy in the final figures.

10. For further information on this topic, portions of several texts are most useful: (1) Christians, et al., *Media Ethics: Cases and Moral Reasoning, 2nd edition* (New York: Longman, Inc., 1987); (2) John L. Hulteng, *The Messenger's Motives: Ethical Problems of the News Media, 2nd edition* (Englewood Cliffs, NJ: Prentice-Hall, 1985); (3) Philip Meyer, *Ethical Journalism* (New York: Longman, Inc., 1987); and (4) Conrad C. Fink, *Media Ethics: In the Newsroom and Beyond* (New York: McGraw-

# 4

# ▼ Special Problems in
# ▼ TV News Ethics

As mentioned earlier, ethical beliefs in broadcasting are usually very individualized and personal. Thus, it is not surprising that reporters often try to find work at news departments where the station's journalistic policies match their own personal beliefs, attitudes, and values. Even when the general match is compatible, however, specific news subjects and events can cause great ethical consternation to the person and station presenting the broadcast. The purpose of this chapter is to discuss four such situations.

Ethical analysis can occur in several different ways; the important point is that there should be some consistency in all decisions, once the basic perspective is adopted. As for applicable philosophical approaches to media ethics, one can look to the ancient Greeks, to nineteenth-century Europeans, or to more contemporary American writers for perspectives on TV news ethics in today's world. While space will not permit a complete investigation of all the philosophical and theological underpinnings of ethical journalistic codes, several approaches should be mentioned, albeit briefly.

## ARISTOTLE AND INTELLECTUAL VIRTUE

Among his many writings about ethics and moral judgments, Aristotle discussed the elusive quality of "right" versus the omnipresent opportunity of "wrong:"

> . . . it is possible to go wrong in more ways than one . . . but there is only one way of being right. That is why going wrong is easy, and going right is difficult; it is easy to miss the bull's eye and difficult to hit it.[1]

But a broadcast professional is highly (and specifically) trained in the world of news gathering; therefore, his or her margin for erroneous information should be substantially reduced compared with that of an average person. Thus, newscasters have the *moral* obligation to be proficient. John Kultgen, in *Ethics and Professionalism*, describes three aspects of professional expertise in technical (and moral) skills:

... First, they vary from profession to profession, since they are determined by the nature of the work ... Second, however, the obligation to acquire appropriate skills is the same for all professionals. It is a *moral* obligation for the professional to be proficient. She must maintain state-of-art competence. She must not practice until she receives the proper training, she must practice only in areas in which she is trained, and she must continue training throughout her career. Being prepared is an obligation of diligence along with working on particular assignments. Third, a profession is an art. Rather than concentrating on a narrow band of concepts under a unified theory, the professional ... must master an eclectic group of concepts tied together only by their relevance to concrete problems. Hence, the professional must have broader intellectual curiosity than most pure scientists ... She must be able to see the universal in the particular as she applies disparate concepts to an endless procession of unique cases.[2]

Obviously, the importance of intellectual virtue is only one aspect of Aristotle's total ethical stance. However, it defines a significant role which TV journalists must accept in today's society.

## UTILITARIANISM AND "THE GREATEST GOOD"

The utilitarian approach is another ethical perspective in media criticism. This philosophy, made popular by Jeremy Bentham and John Stuart Mill in the nineteenth century, suggests that decisions should be made on the basis of maximum interest. For broadcast journalists, this means that all news coverage should be considered for its potential to bring the most reward and cause the least suffering for the viewing public at large. Kultgen elaborates:

Knowledge is one of the most important social goods. It is socially produced in that everyone in society provides material support for inquiry even if he or she does not engage directly in it. What is socially produced should be socially appropriated, so the stock of knowledge is a collective possession of humanity. The stock cannot be appropriated by anyone, but its fruits can be shared. A just society would be organized to share it with everyone equally. Professionalism provides skills based on knowledge that people cannot acquire for themselves, so justice demands that these skills be available to everyone in a way that will contribute equally to the happiness of each.[3]

Thus, the utilitarian viewpoint of the Watergate scandal might have been that the public's right to know about the illegal wiretaps demanded serious investigative reporting by Bob Woodward, Carl Bernstein, and others.

## RAWLS' THEORY OF JUSTICE

Some ethical scholars favor the approach taken by John Rawls in his book, *A Theory of Justice*.[4] Rawls contends that fairness is fundamental to the concept of justice, and that all ethical decisions should be handled with this basic premise in mind. At times, the concept of fairness is presented simply and without controversy. In these cases, according to Christians, Rotzoll, and Fackler:

> . . . fairness means quantity. Everybody in the union doing similar work would fairly receive a 10 percent raise. Teachers would give the same letter grade to everyone with three wrong. All the children at a birthday party should have two cookies. Eliminating arbitrary distinctions expresses [fairness] in its basic sense.[5]

However, as Christians and his colleagues suggest, Rawls has greater difficulty actualizing the concept of fairness in inherently unequal contexts. In these instances, he recommends his "veil of ignorance," i.e., removing all social differentiations if possible. For example, if a large number of people perish in an international plane crash, the news event must be covered. The question is, "how?" Although the victims are from various countries, belong to various cultures, and undoubtedly differ in wealth and political power, should all of them be given equal treatment in the news coverage? According to Rawls' "veil of ignorance," all victims (and their families) should be given balanced air time. In other words, it should not matter that the victims of the crash were American, French or Korean, the story should center upon the entire tragedy. This sentiment also applies to any investigative follow-ups to the cause of the crash.[6]

Regardless of the ethical perspective taken, it is crucial to adapt one's philosophical beliefs to a workable, decision-making structure. Such a mosaic has been put forth by Dr. Ralph Potter of the Harvard Divinity School. "Potter's Box" helps reporters to define the situation, articulate their personal and professional value systems, compare these values with the station's principles, and identify their respective loyalties. Potter's approach to ethics has been discussed in greater detail earlier; however, in this chapter, it will be illustrated through several case studies of ethical dilemmas.[7]

## CASE #1:
## ETHICAL ISSUES CONCERNING NEWS COVERAGE OF TRAGEDY

In the wake of advancing technology, a previously untapped source of influence has evolved in the world of media—the capability for live, on-site reporting. Contemporary viewers now enjoy the luxury of receiving their news on a medium which has the ability to be both a spontaneous and graphic transmitter of societal events. Although the issues deemed newsworthy have not changed, the eyes through which we see these events have become more focused. The new technologies, however, often fail to shield us from the tragic scenes of everyday life. Thus, the line of ethical distinction between delivering the most vivid account of a story and exploiting tragedy becomes hazy at best. At what point does an event become too sensitive to present to the waiting public? And when does standard journalistic reporting overstep the boundaries of decency, sacrificing its ethics for the sake of an enticing segment on the evening news?

Such are the issues worth pondering in light of an event occurring in a midwestern city in August of 1989. On a seemingly normal Friday evening, a man in his late twenties took his wife and children to a nearby river site to watch the boats at night. Before arriving at the landing, he stopped at a nearby convenience store (less than a block away) for cookies and candy to accompany the venture. At the store, some people heard him complain about a leg cramp; however, no one thought much about it at the time.

After leaving the store, the man and his wife made sure the children (ages 8, 4, 2, and 10 months) were safely seatbelted for the short trip to the river parking lot. He started the family station wagon. Then (later claiming that the leg cramp forced him to slam his foot down on the car accelerator), he traveled at a speed of 55 mph down a side street to the river, bolted through a wooden barricade, and plunged into the water as his wife and children screamed in terror. Eyewitnesses told reporters that from their second-story balcony, they had watched the chilling spectacle unfold. And, according to the testimony of these witnesses, the driver made no attempt to stop the car. In fact, he gunned the engine as he sped down the fatal path.

In the confusion that followed, the parents (who were not wearing seatbelts at the time) escaped the sinking vehicle through open windows. They were later rescued by a fisherman who had been cruising around the area in his powerboat. The children, however, were strapped in the car and consequently, could not free themselves. The parents did not try to return to the station wagon for the children, seemingly having suffered from shock and disorientation, but a rescue team from the local sheriff's marine patrol unit arrived within minutes of the accident. They worked for hours to free the children and save their lives. In the end, their attempts were futile; all four children were found floating inside the car.

As the divers pulled the bodies from the river, they found themselves amidst the lights, cameras, and microphones of various local news stations. Some reporters simply recapped the events, choosing not to use the graphic footage taken of the rescue attempt. However, other journalists decided to capitalize on the opportunity for dramatic, live footage of the story. Captured on the six o'clock news, viewers on some stations were shown videotaped shots of the lifeless bodies of four young children being airlifted by helicopters or carried to ambulances. They were also privy to the spontaneous and live interviews of neighbors and relatives of the children.

Based on the unusual number of complaints filed at the various TV stations, it seemed clear that such coverage had exceeded any curiosity for the macabre which may have caused viewers to tune into the broadcast in the first place. Audiences, for the most part, were simply repulsed. They felt that such footage not only breached the standards of ethical reporting, it also placed additional burdens on the loved ones of the four victims.

Using Potter's structural analysis for this case, it is important to:

1. *define* the situation;
2. determine one's own *values* in the matter;
3. compare those values to TV journalistic *principles* and station policies; and
4. decide which *loyalties* are necessary for balanced reporting.

In this particular instance, the situation has been defined as a legitimate news story about four horrible deaths under somewhat suspicious circumstances. Therefore, it is important to bring back some information on the story, especially if it later turns into a murder investigation rather than just a tragic accident. Other factors which must be considered when *defining* the news coverage of this story include the following:

- the technical assistance (e.g., camera, helicopter, etc.) available and necessary for adequate reporting
- the times of day and amount of airtime for the story
- the importance of video/film footage to the story vs. a simple graphic behind the anchor
- the advantages and/or disadvantages of broadcasting "live" from the site, if possible

When asked about the definition of this news story, everyone surveyed agreed that it was a story worthy of broadcast coverage. Donna Skattum (WAYK) typified the responses best by saying:

> ... there seems to be a question as to whether or not the driver of the vehicle attempted to stop. While on the surface, the story appears to be a tragic accident, if the driver didn't even attempt to avoid going into the water, there is a strong possibility the story would change from spot coverage of an accident to murder.

As for other factors in the *definitional* phase of news decision-making, the respondents differed according to the available facilities of their individual stations. They all agreed, however, that if they had all the capabilities mentioned in the questionnaire, they would probably utilize them for this story.

When analyzing the *value* dimension of journalistic ethics, it is important to include both personal and professional beliefs and attitudes. Some of the most significant values which emerge when considering this case are the following:

- an innocent person's right to privacy versus the public right to know
- the responsibility to deliver news as immediately as possible
- the violation of human rights (the man's, his wife's, his children's)

- exploiting a tragic accident
- failure to expose a possible murder
- the responsibility to present the truth in the most factual way possible

Alan Griggs (WSMV) characterized his values on a story like this accordingly:

> I would hope that by the time all the information crossed my desk, it would
> be coming from my reporter on the scene. The story is legitimate. It in-
> volves the deaths of four people with the potential for murder, at least for
> the moment, a major consideration. As with *any* story, it should be handled
> as fairly as possible to reflect the facts.

How should the reporter's values correspond with TV station *principles*? It seems that at least three broadcast rules are apparent in this case:

- information should be aired as immediately as possible
- pictures are the most identifiable feature of televised journalism and should be used often
- truth is always the most important commodity, regardless of picture avail ability

Speaking to some of these principles, the responding news directors had several interesting opinions. Henry Mendoza (KBAK) was staunchly opposed to the growing trend of televised news to adopt slick visuals in favor of good, old-fashioned journalism:

> I believe television does too much LIVE for the sake of using the technol-
> ogy. The LIVE shot should be done only when something is happening
> somewhere at that moment, not as a gimmick.

Donna Skattum (WAYK) was a bit more conservative in her reply, although she, too, felt that the visuals should have a specific purpose if they were to be used at all:

> If video were available, and I were going to air the story, I would use video
> of the scene, police cars and ambulances. I might stop short of using body
> bag shots . . . Television is a visual medium, and the camera is just as much
> a reporting tool as the notebook. At the very least this would be spot cover-
> age of an accident. Perhaps by showing the story we could help prevent
> further accidents at the location. If the story turns out to be more, the foot-
> age of the "accident" scene would be used in future stories.

It was very clear from the responses received on this issue that most re-porters' personal/professional values corresponded favorably to the policies of the stations where they worked. However, when asked to speak more generally about broadcasting graphic violence, the news directors made their *loyalties* very clear. Henry Mendoza (KBAK) deplored the use of violence for its own sake on TV news. As a broadcast journalist, he felt that he owed his viewers more than a lavish display of blood, guts, and gore:

Our ability to go LIVE anywhere, anytime, robs us of the time it should take to try and decipher the context a given situation presents. That occurs often when depicting violent incidents or confrontations. So context needs to be addressed. We should also distinguish between titillation and something the public NEEDS to see. I would always suggest that the question about video be asked, "Do my viewers NEED to see that to understand what happened?" before a decision is made.

Alan Griggs (WSMV) concluded his comments by suggesting a very useful approach to the problem:

One major guideline would be to handle such matters based on the beliefs and mores of the city and surrounding area. We tend to hand such stories more conservatively than stations in much bigger cities. Our writing and production are more straightforward and subdued rather than sensational and graphic. Too many graphic phrases in this business are hackneyed, uncalled for methods of handling a story. They reflect a laziness on the part of the writers involved.

## CASE #2:
## ETHICAL ISSUES CONCERNING
## INTERNATIONAL NEWS COVERAGE

The controversy surrounding the coverage of the drowned children in a midwestern city was heightened, in part, by a violent incident of international scope which had occurred only several days before. The Western world had become reluctantly accustomed to the periodic release of videotapes depicting numerous hostages held captive in the Middle East. It was not, however, prepared for the release of a tape showing a dead man hanging from a rope.

The victim was U.S. Marine Lieutenant Colonel William R. Higgins, 44, who had been abducted the year before in Lebanon while heading an observation team for the UN peacekeeping mission. A pro-Iranian Lebanese Shi'ite organization executed Higgins after their demands for the release of Sheik Abdul Karim Obeid, leader of the Shi'ite Hizballah terrorists, went unheeded. Obeid (and two of his aides) had been captured in a raid by Israelis in southern Lebanon weeks before. The Shi'ite group gave the Israelis a proposed deadline for Obeid's release; if this deadline was not met, they promised to avenge his capture on one or more of their hostages.

The videotape of the hanged hostage, along with a statement typewritten in Arabic, was delivered to news agencies in Beirut several hours after the proposed deadline. The Shi'ites claimed responsibility for Lt. Col. Higgins' death. They also threatened the life of fellow hostage Joseph Cicippio as well, pending the desired release of Obeid by the Israelis.

Unlike other tapes which had clearly indicated a date and time of filming, this footage was surprisingly barren, depicting a head and torso shot of someone claimed to be Higgins, apparently hanging from a rope with his mouth gagged and his hands tied behind his back. The tape also showed a man's legs trussed with a blue cord, and suspended with his shadow swaying against a nearby wall. Because this videotape was the first and only evidence of Higgins' death, the U.S. State Department was hesitant to verify that: (1) Higgins had actually died; and (2) if the body was Higgins', that he had actually died in that place or at that time.

Despite the State Department's declared doubts about the Shi'ite tape, the footage was subsequently acquired and aired by three major news networks: NBC, CBS, and CNN. ABC chose instead to air the still photos of the hanging as opposed to running the video.

The response of many media critics to the graphic depiction of Higgins' hanging on national television was enormously negative. Like most Americans, they voiced shock and concern over the seemingly unlimited boundaries of news presentation. They also questioned the advisability of airing an "unconfirmed" execution by international terrorists, which might ultimately threaten the lives of other hostages as well. To make this latter point even more persuasive, the critics cited 1985's infamous TWA Flight 847 hostage takeover, where terrorists ended up running the news media, not vice versa.

Other media critics took a different perspective, however, presuming that our society was accustomed to a certain amount of violent drama on the air. As such, they found this display in sound keeping with the tragic nature of the event and with the viewer's right to receive the most detailed account possible of any given story.

Which side was "right?" Once again, it is time to use Potter's Box, along with some news directors' responses.

When *defining* the event, it is important to consider that this case was a legitimate news event about the reported assassination of an American military advisor in Lebanon. It also updated the crisis of the hostage situation in the minds of many Americans, which, as noted in an earlier chapter, might prove helpful in ultimately ending the crisis. Further, the tape was available, presentable, and not officially banned by the government. As Steve Snyder (KDLT-TV) stated:

> If there was sufficient evidence of a breach of security if the story was aired immediately, I might be compelled to hold off for a pre-determined amount of time with the stipulation that the story would air as soon as possible.

Next in Potter's configurational scheme is a consideration of the news reporter's personal and professional *values*. Several TV executives were quite candid about their commitments to immediacy, privacy, truth, objectivity, and the viewer's right to know. However, they seemed to make no personal value statements about the hostage crisis in the Mideast. For example, David Russell (WHKE-TV) responded to the question accordingly:

There is no primary information to be gathered by airing this tape that would supersede the family's right to privacy. No necessary information would be withheld from the public by not airing this tape.

Brian Moore (WVGA) was more concerned with the overall issue of human rights than he was with the specific hostage crisis:

It could interfere with the human rights of Lt. Col. Higgins and his family because there was no definite proof of his death. If he was still alive, then his family would have to answer unnecessary questions about his death and it could possibly interfere with his rescue.

Steve Snyder (KDLT-TV), like his other colleagues, was quick to note that his job was to provide news, not to suggest a specific editorial position:

If it is a hoax, it is up to the viewing public to determine its own reaction. It is not my responsibility to censor the news based on whether the story as put out by the terrorists is a hoax. My job is to report events. By reporting this reported assassination, the viewer then has the right and responsibility to determine [his/her] own reaction to the people who perpetrated it.

It appeared from the surveyed responses that each news director's *values* corresponded favorably with his or her station's *principles*. In general, it was felt that this was a network story, and, as such, the local affiliates or independent news operations had little to do with it. Had hostages Higgins and Cicippio come from the local area, there might have been more questions about broadcast coverage of the actual execution and subsequent threat of execution. As it was, there were no conflicts between the news directors' *values*, the station's *principles*, or the *loyalties* which would necessarily have been taken by the reporters.

## CASE #3:
## ETHICAL ISSUES CONCERNING NEWS CELEBRITIES IN THE PUBLIC EYE

When considering the wide range of stories which become newsworthy on a daily basis, one must include the myriad of seemingly unimportant incidents involving famous American people. Although these celebrity scoops lack significance on a higher level, there is no denying the relative merit they hold in satisfying the insatiable curiosity of a culture with no royalty about which they can marvel and worship.

The figures in the news world are certainly no exception. Even the local television weatherman is often the subject of conversation at dinner tables throughout mainstream America. A difficulty arises, however, when one attempts to draw a line of distinction between the public and private worlds of celebrities in relation to the audience's right and desire to know about them.

Does fame necessarily force the individual to become a permanent fixture of the public domain? On the other hand, as an influential figure in society, does the celebrity have a responsibility to remain clearly visible under the public eye? Specifically, does the public deserve to know all?

A poignant example of this ethical dilemma occurred in July 1989, when a local television news celebrity dodged the public while coming to terms with personal problems. This particular newscaster, who had dominated the local news scene for more than 20 years, sought treatment for alcoholism in a West Coast hospital. The only statements released by the broadcaster seemed indicative of a sincere attempt to tackle the disease.

The controversy arose because viewers had noticed the broadcaster's absence for several weeks from the local evening news without comment from the station management or other anchorpeople. Once the story was disclosed to the public, the station was barraged with questions concerning the secrecy of the matter and the unusual length of time which had elapsed before releasing the story. The station manager stated that keeping the information undisclosed was the request of the broadcaster, who seemed to feel that the ordeal was ultimately a private matter. Rumor immediately spread, however, that the information was withheld from the public because the broadcaster's absence coincided with a ratings sweep period for the station. Admittedly, the delayed announcement of the news anchor's entry into the treatment center seemed directly related to the extremely high ratings during that time.

The pertinent questions raised in this case deal with the ethical treatment of a local news celebrity. Did the public have a right to know the personal difficulties of the newscaster? And if the public should have been privy to this information, was the question of immediacy a relevant one?

In *defining* this news event, all persons surveyed agreed that it was a legitimate story which required press coverage. They also felt that this news item should be released in a timely fashion or the station might risk a loss of credibility to its viewership. The possibilities of news dissemination ranged from a simple press release to a live news conference.

John McKean (KGGM) was most candid in his opinion that news organizations could *not* afford to taint their reputations by repressing station news:

> It is vitally important for news organizations to avoid the charge of unfairness and hypocrisy. They must be as open and forthcoming as they expect public officials and public figures to be.

Dick Nelson (KCTV) seemed a bit more liberal in his time framework, but he agreed nonetheless that some information should be released within a relatively short period:

> It's not a matter of great urgency, but the public has a right to know about a public figure's performance, particularly as it might affect the performance of his public "duties"—in this case, reporting news.

Once again, the professional *value* of individual privacy versus the public's right to know was the most important consideration for these news directors. They also discussed the price one must pay for the glory of being a celebrity. Finally, equal treatment of all newsmakers was essential—to ignore a station employee's behavior would be the worst form of bias these news directors could create.

Bob Reichblum (WJLA) was most adamant in his reply:

> You can't have two standards—important developments affecting you or others must be treated equally.

John McKean (KGGM) was equally strong in his opinion, saying that newscasters should not necessarily enjoy the rights of private citizenship:

> Broadcasters do make policy—they make judgments about what is news. They are public performers. They cannot pretend to be private citizens unaccountable to the public they purport to serve.

Dick Nelson (KCTV) summarized his colleagues' comments, by stating:

> In my opinion, a journalist is a kind of unelected public official who should be answerable for anything which directly or indirectly relates to the performance of his job.

When the news directors related their personal *value* orientation to the station's *principles* and policies, the major issue seemed to be one of immediacy in releasing the information to the public. No one seemed to see the station's reticence of the news anchor's whereabouts as an attempt to boost the July ratings sweep period; rather, each executive saw the delay as finding the correct balance between respecting an individual's privacy and honoring the public's right to know. Thus, the news directors felt no *loyalty* dilemma in this particular case. As Dick Nelson (KCTV) stated:

> The public's right to know has to be balanced against the person's right of privacy. *Delaying* release of this information is an acceptable balance. *Withholding* it would not be.

## CASE #4:
## ETHICAL ISSUES CONCERNING THE RE-CREATION OF NEWS EVENTS

A prominent characteristic of the contemporary news media is its ability to present newsworthy events with such spontaneity that the viewer has a ringside seat to community, national, and global occurrences as they happen. While television generally upholds such a distinguished reputation with consistency, there are situations in which reporters and cameras are one step behind the news. The recent technological solution to such instances, however, has inadvertently created an interesting ethical dilemma.

Some TV news stations, failing to capture certain events authentically, have resorted to *simulation devices* in order to portray the original flavor of the story to their audiences. Although the desired effect is often achieved, the consequences in a larger context are questionable at best. This is because both the subject of the story and the viewer can leave themselves vulnerable to the slight but often significant changes that occur with an adaptation of an event. In other words, one is often left to interpret the network or station's adaptation as well as the event itself. Needless to say, such circumstances could open up a "Pandora's box" for all concerned.

A primary example of simulated adaptation and its problems occurred in July of 1989, when U.S. diplomat Felix S. Bloch emerged as a focus of international curiosity. He was under surveillance at the time by the Federal Bureau of Investigation, which had acted upon certain allegations that Bloch had been spying for the Soviet Union. Although the details were murky at best, ABC News took the initiative and re-created a crucial element in the evidence against Bloch. During the 6:30 and 7:00 P.M. programs, "World News Tonight" ran a tape with the story, showing what appeared to be an American diplomat handing a briefcase to a Soviet agent. The diplomat pictured was portrayed as Bloch, and the footage on the tape was supposedly very similar to a tape possessed by the Federal Bureau of Investigation. Unfortunately, however, the network failed to inform its viewing public during the broadcast that the segment was a simulation of what was alleged to be the real event.

ABC had actually used members of its own staff to play the roles of Bloch and the Soviet agent. The videotape was then processed through a new technological device which gives the illusion of hidden camera footage. While the process itself is not uncommon, the resulting simulation is always labeled as such when presented on the air. The Bloch simulation, however, was sent during the 6:30 P.M. feed without the word *simulation* on the screen. As a result, those viewing the 1-minute segment believed the feed to be actual footage of an exchange between Bloch and the Soviets. ABC ran a corrected version of the tape during the 7:00 P.M. show, but most of the ABC affiliates had already used the earlier feed in their evening broadcast.

To be sure, accidents can occur with some frequency in an industry that holds *immediacy* as its credo. However, TV broadcasters should always be aware that simulating news events often compromises both the integrity of newsworthy subjects as well as the vulnerable impression of television's viewing public.

When applying Potter's Box to this situation, the *definition* of the event is clear. Broadcast executives would certainly consider this a legitimate news event, although the timing of the presentation might be a bit precipitous. In July 1989, Bloch was simply being investigated; no formal charges were handed out, and the simulation of a possible briefcase exchange was premature at best. Still, the notion of yet another international spy in collusion with the Soviets would definitely make for good news copy.

As for the journalist's *values* in this case, all of the respondents were horrified at the methods used by ABC News in presenting this story. However, there seemed to be some difference of opinion about the general usefulness of simulation devices. Some argued that simulations were desirable, and even necessary in some cases. Liz Talbot (WVTV), for example, noted that a simulated sequence can enhance a visual presentation at the scene of the crime. However, as she stated, there should definitely be some presentational restrictions:

> Using camera moves and reporter standups, along with eyewitness accounts (but *not* actors), it can be done accurately and professionally.

Henry Mendoza (KBAK), on the other hand, was adamantly opposed to any simulations at all:

> ABC's use of reenactments, and subsequent explanation of how and why it was done, raised questions which should NEVER even come up. The notion that any news organization could even consider reenacting a news event is abhorrent and has brought news into an area it should NEVER have entered.

David Russell (WHKE) added to Mendoza's point by ending with this fitting comment:

> The organization that the people trust to deliver the news in an unbiased fashion should not be in the business of interpreting the news.

Clearly, these people (and others like them) might have run into difficulties with the *principles* in ABC's policies to air such a story without labeling it as a reenactment. And, in the final analysis, they would have faced an ethical dilemma between a *loyalty* to the network . . . or to their own personal/professional values.

These four case studies serve to underscore the fact that while general policies and ethical codes in TV journalism can be devised, one must always consider the specific situation and its relationship to the business of broadcasting before deciding whether or not it should be aired. And, after that decision has been made, a more difficult judgment arises: What resources and time commitments should be made to the story?

Unfortunately, these considerations have not grown any easier over the last several decades. In fact, the development of new technologies, along with the threat of further economic restraints, shrinking audiences, and governmental roadblocks, will make the TV news director even more besieged in the 1990s and beyond. The next chapter addresses future developments . . . and future problems in broadcast journalism.

## Notes

1. Aristotle, *Nichomachean Ethics*, trans. J.A.K. Thompson (London: George Allen & Unwin, 1953), Book II, p. 66.

2. John Kultgen, *Ethics and Professionalism* (Philadelphia: University of Pennsylvania Press, 1988), pp. 356-357.

3. *Ibid.*, p. 194.

4. John Rawls, *A Theory Justice* (Cambridge: Harvard University Belknap Press, 1971).

5. Christians, et al., p. 14.

6. Incidentally, American media coverage of international tragedies such as the hostage crisis and terrorist bombings have come under close scrutiny in recent years. One of the sharpest criticisms against TV news is its propensity to highlight events that only include Americans; U.S. broadcast coverage generally ignores citizens from other countries. Some critics claim that this "exclusive" media attention actually encourages terrorist attacks against Americans because it is an effective way to gain notoriety.

7. These case studies have been compiled with the help of my research assistant, Charles E. Morris III, at Boston College.

# 5

# TV Ethics and the Future

The previous chapters have explored some inherent limitations of television news from several different points of view. Historically, for example, radio news clearly created a legacy of dramatic commentary, celebrity status, and popular culture, all of which eventually found equal acceptance in televised journalism. Economically, the survival of TV news continues to exist through monies from advertising revenue; thus, the fear of losing a sponsor through controversial news coverage must be forever present in the minds of news directors. Technologically, electronic journalism has developed its reputation of immediacy and visual clarity through the constant acquisition of state-of-the-art equipment; to sacrifice these components in favor of presenting more in-depth news coverage (but less visual and instantaneous) becomes an overwhelmingly unpopular idea among competing commercial stations and networks.

With this in mind, television news directors often find themselves dealing with the challenge of maintaining a delicate balance between providing adequate information to the public and keeping a commercial enterprise healthy and happy. Some of these ethical dilemmas have been discussed in Chapters 3 and 4 of this text; however, other problems—especially those dealing with future economic, technological, and governmental developments—are still evolving. The purpose of this chapter is to acquaint the reader with several of these issues.

## CONSUMER CAMCORDERS

In a 1989 documentary produced by Ted Koppel and ABC News, Koppel commented upon the growing amount of media technology which has become available to the general public. He said:

> The world is in the early stages of a revolution that it has barely begun to understand. We think we know something about the impact of television because it's been with us for fifty years. But recently, television has begun falling into the hands of the people.[1]

In part, Koppel's statement was directed toward the "revolution" of home video and its growing range of possibilities.

In 1989, 7,500,000 camcorders were sold worldwide,[2] providing millions of people with the opportunity to capture important moments on videotape. Some of these "moments" are intensely personal, but an increasing number of videophiles are using consumer camcorders to provide news footage for local stations. Recent examples of such home-grown footage include the following:

- A highly publicized tape showing a babysitter abusing a small infant while the parents were away at work
- A drug raid covered by one of the arresting police officers who attached the video camera to his helmet
- Dramatic footage of the explosion on the USS Iowa
- The collapse of a tower built for Pope John Paul II's arrival in Texas
- The West German air show plane collision
- A tornado in the Southwest blowing a car off the ground
- Horrifying footage of the collapse of the Oakland Bay Bridge during the October 1989 San Francisco earthquake

Most news directors applaud the availability of home video technology; they contend that it makes their job much easier. In fact, Steve Ridge, a vice-president of Frank N. Magid Associates,[3] estimates that about 20% of all network-affiliated stations depend upon some in put of viewer home video footage.[4] And, to many broadcast executives, the technology may be new, but the concept is as old as radio. As Marv Rockford (KCNC-TV, Denver) says:

> Having people who are not a part of your regular news staff, who just supply you with news material whether they're pictures or words or information—it was simply taking that old idea of the news stringer and marrying it to the realization, the recognition, that there is an awful lot of that home video equipment out there.[5]

Jeff Klotzman (WKOW-TV, Madison WI) agrees with Rockford, commenting that, in the business of TV news:

> You can never have enough eyes and ears out there. It may not be the greatest quality video but in a breaking news story it's better than nothing.[6]

Most of the time, stations pay relatively little for video footage—between $25 and $100[7]; however, the major reward for most amateur camera people is the glory of being identified as a contributor to that day's newscast. Paul Garrin, an amateur videographer who shot some dramatic footage of a New York neighborhood riot, asserts:

> The video medium, or television, has become now democratic in a way that it never has been in our history. And now, anybody on the street can catch something just at a whim or by chance. And I think that that's a great asset to the news networks.[8]

Despite the seeming popularity of home video contributions to TV news, however, several industry professionals can foresee serious ethical dilemmas if camcorder footage is overused. First of all, one must always consider carefully *who* shot the videotape, and *why* they may have taken their camera along to the news event. Everette Dennis from the Gannett Center for Media Studies amplifies upon this point:

> We don't always know the motivation of the people who are taking the pictures. If somebody's doing it for a special interest cause, it's very different than a network newsperson who's trying to be impartial in their treatment of an issue.[9]

For example, some political, religious, or environmental activist may actually stage an event, and record it on videotape to further his or her cause. News directors, if nondiscriminatory in their picture selection, might choose the story for its visual value rather than questioning its story value. In such cases, the TV station — and its viewing audience — would be duped by a manipulative perpetrator.

Another ethical dilemma confronting the overuse of home video footage involves the employment security of professional camera people. In some stations, staff camera crews refer to amateur productions as "scab tapes," and have persuaded their unions to demand agreements which restrict their use. One such example exists at WCAU-TV in Philadelphia, where *TV Guide* reports:ß

> ... the station is allowed to use only tape that its own crews would not have been able to shoot (or amateur footage that's clearly better than the station's own); use is restricted to one minute; and amateurs cannot provide interviews.[10]

The most important ethical consideration of home video usage on TV news, however, is its potential to sacrifice the *accuracy* of a story in favor of immediacy. As has been previously established in Chapter 3 of this text, broadcast executives value *accuracy* more highly than any other factor of good journalism. Unfortunately, accuracy can sometimes be compromised for the need to be first with a story, and the potential danger of factual error multiplies with the use of amateur video footage. The most glaring example of unsubstantiated video happened in the Soviet Union a few years ago. An amateur videographer submitted footage of what he claimed to be a fire emanating from a chemical plant in Chernobyl; the videotape was later revealed to be shot near a cement factory in Italy.

Don Browne, the executive director for NBC News, describes the implicit problems of presenting a news event too quickly:

> There is a danger that because of the speed and demand and competitiveness that some of those things can slip through the cracks. If we rush things on the air without checking them out because of competitive pressures or because we have too much information to deal with, we're going to have credibility problems.[11]

Undoubtedly, the newfound availability of amateur video footage can make news broadcasts more dramatically exciting at times. However, in the future, the TV news director will most certainly be called upon to make ethical decisions regarding the continued use of such visuals.

## ACCESSIBILITY

The presence of home camcorders in electronic journalism is only one instance of growing consumer accessibility to technology. And in a world where technological devices are becoming increasingly available, almost anyone has the capability to disseminate information. Consider, for example, the reaction of American Sandinista sympathizers when Daniel Ortega was denied a visa to come to the United States in the spring of 1989. Without the help of a network or local station, this group was actually able to rent a studio, hire a crew, work from a satellite ground station in Managua, and broadcast 45 minutes of airtime to interested Americans via satellite. The total cost was less than $7500.[12]

The opportunity to televise events in this manner is a credit to technological innovation. It is especially important to note this achievement when remembering that 50 years ago, television was seen as little more than an exciting experimental challenge.

The development of affordable, new technologies has done far more than supply consumers with equipment and software, however. It has enabled other professional news organizations to enter into the television marketplace, thus penetrating previous network domination in information dissemination. Today, viewers can choose from a menu of information alternatives; no longer are they restricted to news coverage solely provided by NBC, CBS, or ABC.

Several reputable TV journalistic services have emerged in the past few years; among these, the most critically acclaimed organization has been the Cable News Network. Since its first broadcast in June 1980, CNN has risen dramatically in stature, having hired extremely professional journalists as well as making sure that no one arrives on the scene before the CNN remote vans. Ed Turner, an executive at the Cable News Network, describes the very simple philosophy behind its news operation:

> We are today what radio once was in the form of the BBC's world service and VOA and its counterparts. We have moved, and I think are supplanting, that radio service now with television — television news.[13]

Needless to say, CNN's incredible decade of success has not met with much enthusiasm at ABC, CBS, or NBC. And local affiliates are even more disturbed by its presence. As Donald Varyu (KING-TV, Seattle) observes:

> For a long time, the ultimate solution in the struggle with cable has been local news. There is no way CNN can give Seattle news to Seattle or Los

Angeles news to Los Angeles. But now, some stations are making ar-
rangements to put local news inserts on CNN. That's the first step on the
road to erasing our advantage.[14]

As noted in an earlier chapter, network news viewership has fallen con-
sistently since 1979. More competition in news programming will translate
into even lower ratings and shares for broadcast networks in the future. In
addition, according to Edwin Newman, CBS, ABC, NBC, and their affiliates
are now faced with the difficult task of finding a significant place for themselves among a crowd:

> More and more local stations and services like the Cable News Network
> are able to trot around the globe just as ABC, CBS, and NBC can. Even-
> tually, what may set the networks apart is their ability to provide reliable,
> authoritative reporting on national and international affairs. TV news
> budget cutbacks are another reason that longer, deeper pieces and com-
> mentary may become more plentiful—they're cheaper to produce.[15]

Longer editorial commentary packages do exist in network news broadcasts; however, another trend has become even more popular . . . and alarm-
ing. This is the growing tendency for news presentations (at both network and
local levels) to incorporate more show business techniques than journalistic
information. At the 1989 Radio-Television News Directors Association Convention in New York, broadcast executives voiced great concern about com-
promising their professional values in order to lessen audience erosion.
Colleen Dudgeon (WBBM, Chicago) made her anxieties about the situation
well known:

> There's a continuing pressure to blur the lines between news and entertain-
> ment as well as news and drama. We have to keep news separate and
> distinct from other aspects of the TV industry. We're in the business of
> covering news, not in the business of being entertainment. The whole idea
> that we can hire actors and actresses to portray a news story worries me.[16]

Unfortunately, however, a current fact in TV journalistic life is that more
viewers seem to enjoy entertainment-oriented news programming than the
more traditional forms of information presentation. Shows like "Hard Copy,"
"Inside Edition," and "A Current Affair" seem to capture high ratings num-
bers; thus, the news director must face the current and future ethical dilemma
of "walking the tightrope between news and entertainment."[17]

## COMPUTERS AND JOURNALISTIC ETHICS

The decade of the 1980s has often been referred to as "the age of information," due in part to the meteoric rise in computer sophistication as well as
to the growing affordability and availability of most of the new technologies.

However, these innovative methods of information storage and retrieval are not proving to be as advantageous to TV journalists as media critics had once hoped. While it is true that computer-generated data can multiply the resources most reporters use in news gathering, it is also true that access to electronic records has become difficult, tedious, and sometimes legally impossible.

First of all, it is important to note that the Freedom of Information Act (FOIA) does not include entry into federal computer data at this time. This is due to the fact that the FOIA admittedly needs to be updated with regard to rapidly developing changes in technology. However, even after the language in this Act has been modified to reflect current conditions, the difficulties in obtaining access to information will still exist. Scott Armstrong, executive director for the National Security Archive, has alluded to four such major problems which may occur more often in the future:

1. The destruction of records in an electronic environment is easier than before. A single key stroke can eliminate years of data.
2. The development of electronic databases has become yet another excuse for increasing classification and other controls on federal information.
3. Electronic information poses new problems in terms of fees to be charged for access or copies.
4. Public interest advocates often have to battle private information industries for government dissemination of information.[18]

Thus, at the very least, new technological developments in creating and maintaining computerized records have the potential to compromise basic journalistic standards of immediacy, objectivity, and accuracy.

The above commentary may seem unrealistic, overreactive, and Orwellian to some readers; however, there is substantial evidence to back these claims. Consider, for example, the recent experiences of *The Boston Globe* reporter Peter Gosselin.[19] While investigating several stories for his newspaper, Gosselin ran into various roadblocks, including the following:

- Because of a general lack of information among government public relations officials and freedom of information officers, Gosselin was unable to receive an adequate understanding of his rights under the FOIA. In fact, some government employees seemed totally unaware of the agency's computer ability to store data on magnetic tapes, on floppy disks, or by other electronic means.

- The time spent in receiving data releases was excruciatingly long. According to Gosselin, news organizations and agency officials may take months to negotiate which records will be made available. Most of the problems involve issues of privacy.

- Relatedly, the federal government does not seem to have an overall policy concerning the dissemination of electronic data. As a result, reporters who ask for information from one agency often must wait for a decision from another one. Even worse, not all governmental agencies agree on data dissemination; therefore, one agency may agree to the information request, while another may deny it.

- Gosselin also found that computer software is seriously outdated in some federal agencies. Thus, even if a journalist is authorized access to the desired data, he or she may have trouble retrieving needed information.

- Some of the government's information can be unreliable and inaccurate. Gosselin discovered this shortcoming while investigating a money laundering scandal for *The Boston Globe*. Apparently a government clerk at a computer center had randomly added zeros to some cash figures; the resulting discrepancies amounted to $59 billion.

The computer-caused complexities found in the Freedom of Information Act underscore some of the difficulties facing broadcast journalists in their present and future news gathering efforts. Unfortunately, however, these problems are occurring during the same time TV news departments are cutting budgetary expenditures for investigative reporting, consumer response lines, etc. The ethical quandaries abounding from such combined circumstances will plague broadcast news directors in the 1990s and beyond. Unless more monies are devoted to news production, reporters will be forced to rely upon questionable information at best. Thus, the "accuracy versus immediacy" dilemma will continue to plague TV journalists at both the network and local levels.

The ethical concerns discussed in this chapter are by no means exhaustive. However, they do illustrate the point that journalists and news directors will continue to face moral dilemmas based upon the inherent limitations within the medium of television—time, economics, technology, governmental influence, advertising support, and viewer popularity. Such factors were inherited by TV newscasters, and such will be their legacy.

Therefore, the approach to ethical problem-solving in electronic journalism is not *absolute*; instead, it is *relational* to the capitalistic structure of free enterprise as well as to the commitment of serving the public interest. As such, broadcast codes of ethics should be read in the spirit in which they were written, not as rigid moral dictates. As Casey Bukro has noted:

> The [SPJ/SDX] code of ethics is what it was intended to be: Guidelines
> and a statement of principles to usher us in the direction of responsible
> journalism, accuracy, objectivity and fair play. It is not intended to be a
> substitute for commonsense, good taste or careful deliberation that can steer
> us safely through the maze of complex and often emotional issues that typify

ethics . . . . The code was never meant to be a club or a straightjacket, which would be resented quickly by journalists known for their independence. It's intended to make us think about the way we perform professionally, while at the same time offering enough specifics to say: "Beware!"[20]

Unfortunately, the current state of TV ethics mirrors much of today's world. However, broadcast professionals have the opportunity — and the responsibility — to take the initiative in promoting ethical standards in news presentation. It's not always easy; but it is most necessary, as Donna Skattum (WAYK) observes:

Just as there has been a gradual erosion of ethics in society, there has been a decline in journalism as well. I don't think it's irreparable, but it's going to take a concerted effort by all journalists to stem the tide. As long as people continue to be concerned with getting "the story," with being "first" no matter what the cost, the problems will remain, and multiply.[21]

Finally, ethical codes should be updated continually to address technological, economic, political, and societal changes. While it is true that these codes cannot possibly address all the specific problems one encounters at a news director's desk, they do serve as an excellent blueprint for the ethical decision-making process. Whether one follows RTNDA guidelines, SPJ/SDX tenets, or a station policy based upon a sophisticated Potter's Box, the hopeful result will be better journalism and better public service.

## Notes

1. "The Koppel Report: Television—Revolution in a Box," produced by ABC News (New York), 1989.

2. *Ibid.*

3. Frank N. Magid Associates, located in Dubuque, IA, is a leading media consulting organization.

4. Joanna Elm, "Tonight's Hot Story is Brought to You...By You!" *TV Guide* (February 24, 1990), p. 25.

5. "The Koppel Report: Television—Revolution in a Box."

6. Elm, p. 25.

7. *Ibid.*

8. "The Koppel Report: Television—Revolution in a Box."

9. *Ibid.*

10. Elm, p. 26.

11. "The Koppel Report: Television—Revolution in a Box."

12. *Ibid.*

13. *Ibid.*

14. Ralph Tyler, "News Business vs. Show Business is on News Directors' Minds," *Variety* (September 6, 1989), p. 67.

15. "Television: The Power of Pictures," 1988.

16. Tyler, p. 67.

17. *Ibid.*

18.  Abel Montez, "Computer Age Poses New Obstacles, New Options," *Computer Access: Maze or Miracle* (Chicago, IL:  Society of Professional Journalists' Freedom of Information Committee, 1989-1990), p. 3.

19.  *Ibid.*, p. 4.

20.  Casey Bukro, "The Code is Intended to Make Us Think Professionally," in *Solutions Today for Ethics Problems Tomorrow* ( Chicago, IL:  The Ethics Committee of the Society of Professional Journalists, 1989-1990), p. 22.

21.  Ethics survey, November 1989.

## Appendix A

# Television News Coverage: Network Censorship

by Spiro T. Agnew,
Vice President of the United States

*Delivered at the Mid-West Regional Republican*
*Committee, Des Moines, Iowa, November 13, 1969*

Tonight I want to discuss the importance of the television news medium to the American people. No nation depends more on the intelligent judgment of its citizens. No medium has a more profound influence over public opinion. Nowhere in our system are there fewer checks on vast power. So, nowhere should there be more conscientious responsibility exercised than by the news media. The question is, Are we demanding enough of our televised news presentations? And are the men of this medium demanding enough of themselves?

Monday night a week ago, President Nixon delivered the most important address of his Administration, one of the most important of our decade. His subject was Vietnam. His hope was to rally the American people to see the conflict through to a lasting and just peace in the Pacific. For 32 minutes, he reasoned with a nation that has suffered almost a third of a million casualties in the longest war in its history.

When the President completed his address—an address, incidentally, that he spent weeks in the preparation of—his words and policies were subjected to instant analysis and querulous criticism. The audience of 70 million Americans gathered to hear the President of the United States was inherited by a small band of network commentators and self-appointed analysts, the majority of whom expressed in one way or another their hostility to what he had to say.

It was obvious that their minds were made up in advance. Those who recall the fumbling and groping that followed President Johnson's dramatic disclosure of his intention not to seek another term have seen these men in a genuine state of nonpreparedness. This was not it.

One commentator twice contradicted the President's statement about the exchange of correspondence with Ho Chi Minh. Another challenged the

President's abilities as a politician. A third asserted that the President was following a Pentagon line. Others, by the expression on their faces, the tone of their questions and the sarcasm of their response made clear their sharp disapproval.

To guarantee in advance that the President's plea for national unity would be challenged, one network trotted out Averell Harriman for the occasion. Throughout the President's message, he waited in the wings. When the President concluded, Mr. Harriman recited perfectly. He attacked the Thieu Government as unrepresentative; he criticized the President's speech for various deficiencies; he twice issued a call to the Senate Foreign Relations Committee to debate Vietnam once again; he stated his belief that the Vietcong or North Vietnamese did not really want military take-over in South Vietnam; and he told a little anecdote about a "very, very responsible" fellow he had met in the North Vietnamese delegation.

All in all, Mr. Harriman offered a broad range of gratuitous advice challenging and contradicting the policies outlined by the President of the United States. Where the President had issued a call for unity, Mr. Harriman was encouraging the country not to listen to him.

A word about Mr. Harriman. For 10 months he was America's chief negotiator at the Paris peace talks—a period in which the United States swapped some of the greatest military concessions in the history of warfare for an enemy agreement on the shape of the bargaining table. Like Coleridge's Ancient Mariner, Mr. Harriman seems to be under some heavy compulsion to justify his failure to anyone who will listen. And the networks have shown themselves willing to give him all the airtime he desires.

Now every American has a right to disagree with the President of the United States and to express publicly that disagreement. But the President of the United States has a right to communicate directly with the people who elected him, and the people of this country have the right to make up their own minds and form their own opinions about a Presidential address without having a President's words and thoughts characterized through the prejudices of hostile critics before they can even be digested.

When Winston Churchill rallied public opinion to stay the course against Hitler's Germany, he didn't have to contend with a gaggle of commentators raising doubts about whether he was reading public opinion right, or whether Britain had the stamina to see the war through.

When President Kennedy rallied the nation in the Cuban missile crisis, his address to the people was not chewed over by a roundtable of critics who disparaged the course of action he'd asked America to follow.

The purpose of my remarks tonight is to focus your attention on this little group of men who not only enjoy a right of instant rebuttal to every Presidential address, but, more importantly, wield a free hand in selecting, presenting and interpreting the great issues in our nations.

First, let's define that power. At least 40 million Americans every night, it's estimated, watch the network news. Seven million of them view ABC, the remainder being divided between NBC and CBS.

According to Harris polls and other studies, for millions of Americans the networks are the sole source of national and world news. In Will Rogers' observation, what you knew was what you read in the newspaper. Today for growing millions of Americans, it's what they see and hear on their television sets.

Now, how is this network news determined? A small group of men, numbering perhaps no more than a dozen anchormen, commentators and executive producers, settle upon the 20 minutes or so of film and commentary that's to reach the public. This selection is made from the 90 to 180 minutes that may be available. Their powers of choice are broad.

They decide what 40 to 50 million Americans will learn of the day's events in the nation and in the world.

We cannot measure this power and influence by the traditional democratic standards, for these men can create national issues overnight.

They can make or break by their coverage and commentary a moratorium on the war.

They can elevate men from obscurity to national prominence within a week. They can reward some politicians with national exposure and ignore others.

For millions of Americans the network reporter who covers a continuing issue—like the ABM or civil rights—becomes, in effect, the presiding judge in a national trial by jury.

It must be recognized that the networks have made important contributions to the national knowledge—for news, documentaries and specials. They have often used their power constructively and creatively to awaken the public conscience to critical problems. The networks made hunger and black lung disease national issues overnight. The TV networks have done what no other medium could have done in terms of dramatizing the horrors of war. The networks have tackled our most difficult social problems with a directness and an immediacy that's the gift of their medium. They focus the nation's attention on its environmental abuses—on pollution in the Great Lakes and the threatened ecology of the Everglades.

But it was also the networks that elevated Stokely Carmichael and George Lincoln Rockwell from obscurity to national prominence.

Nor is their power confined to the substantive. A raised eyebrow, an inflection of the voice, a caustic remark dropped in the middle of a broadcast can raise doubts in a million minds about the veracity of a public official or the wisdom of a Government policy.

One Federal Communications Commissioner considers the powers of the networks equal to that of local, state and Federal Governments all combined.

Certainly it represents a concentration of power over American public opinion unknown in history.

Now what do Americans know of the men who wield this power? Of the men who produce and direct the network news, the nation knows practically nothing. Of the commentators, most Americans know little other than that they reflect an urbane and assured presence seemingly well-informed on every important matter.

We do know that to a man these commentators and producers live and work in the geographical and intellectual confines of Washington, D.C., or New York City, the latter of which James Reston terms the most unrepresentative community in the entire United States.

Both communities bask in their own provincialism, their own parochialism.

We can deduce that these men read the same newspapers. They draw their political and social views from the same sources. Worse, they talk constantly to one another, thereby providing artificial reinforcement to their shared viewpoints.

Do they allow their biases to influence the selection and presentation of the news? David Brinkley states objectivity is impossible to normal human behavior. Rather, he says, we should strive for fairness.

Another anchorman on a network news show contends, and I quote: "You can't expunge all your private convictions just because you sit in a seat like this and a camera starts to stare at you. I think your program has to reflect what your basic feelings are. I'll plead guilty to that."

Less than a week before the 1968 election, this same commentator charged that President Nixon's campaign commitments were no more durable than campaign balloons. He claimed that, were it not for the fear of hostile reaction, Richard Nixon would be giving into, and I quote him exactly, "his natural instinct to smash the enemy with a club or go after him with a meat axe."

Had this slander been made by one political candidate about another, it would have been dismissed by most commentators as a partisan attack. But this attack emanated from the privileged sanctuary of a network studio and therefore had the apparent dignity of an objective statement.

The American people would rightly not tolerate this concentration of power in Government.

Is it not fair and relevant to question its concentration in the hands of a tiny, enclosed fraternity of privileged men elected by no one and enjoying a monopoly sanctioned and licensed by Government?

The views of the majority of this fraternity do not—and I repeat, not—represent the views of America.

That is why such a great gulf existed between how the nation received the President's address and how the networks reviewed it.

Not only did the country receive the President's address more warmly than the networks, but so also did the Congress of the United States.

Yesterday, the President was notified that 300 individual Congressmen and

50 Senators of both parties had endorsed his efforts for peace.

As with other American institutions, perhaps it is time that the networks were made more responsive to the views of the nation and more responsible to the people they serve.

Now I want to make myself perfectly clear. I'm not asking for Government censorship or any kind of censorship. I'm asking whether a form of censorship already exists when the news that 40 million Americans receive each night is determined by a handful of men responsible only to their corporate employers and is filtered through a handful of commentators who admit to their own set of biases.

The questions I'm raising here tonight should have been raised by others long ago. They should have been raised by those Americans who have traditionally considered the preservation of freedom of speech and freedom of the press their special provinces of responsibility.

They should have been raised by those Americans who share the view of the late Justice Learned Hand that right conclusions are more likely to be gathered out of a multitude of tongues than through any kind of authoritative selection.

Advocates for the networks have claimed a First Amendment right to the same unlimited freedoms held by the great newspapers of America.

(But the situations are not identical. Where *The New York Times* reaches 800,000 people, NBC reaches 20 times that number on its evening news. {The average weekday circulation of the *Times* in October was 1,012,367; the average Sunday circulation was 1,523,558.} Nor can the tremendous impact of seeing television film and hearing commentary be compared with reading the printed page.)

A decade ago, before the network news acquired such dominance over public opinion, Walter Lippman spoke to the issue. He said there's an essential and radical difference between television and printing. The three or four competing television stations control virtually all that can be received over the air by ordinary television sets. But besides the mass circulation dailies, there are weeklies, monthlies, out-of-town newspapers and books. If a man doesn't like his newspaper, he can read another from out of town or wait for a weekly news magazine. It's not ideal, but it's infinitely better than the situation in television.

There if a man doesn't like what the networks are showing, all he can do is turn them off and listen to a phonograph. Networks, he stated, which are few in number, have a virtual monopoly of a whole media of communications.

The newspapers of mass circulation have no monopoly on the medium of print.

Now a virtual monopoly of a whole medium of communication is not something that democratic people should blindly ignore. And we are not going to cut off our television sets and listen to the phonograph just because the airways belong to the networks. They don't. They belong to the people.

As Justice Byron White wrote in his landmark opinion six months ago, it's the right of the viewers and listeners, not the right of the broadcasters, which is paramount.

Now it's argued that this power presents no danger in the hands of those who have used it responsibly. But, as to whether or not the networks have abused the power they enjoy, let us call as our first witness former Vice President Humphrey and the city of Chicago. According to Theodore White, television's intercutting of the film from the streets of Chicago with the current proceedings on the floor of the convention created the most striking and false political picture of 1968—the nomination of a man for the American Presidency by the brutality and violence of merciless police.

If we are to believe a recent report of the House of Representatives Commerce Committee, then television's presentation of the violence in the streets worked an injustice on the reputation of the Chicago police. According to the committee findings, one network in particular presented, and I quote, "a one-sided picture which in large measure exonerates the demonstrators and protesters." Film of provocations of police that was available never saw the light of day, while the film of a police response which the protesters provoked was shown to millions.

Another network showed virtually the same scene of violence from three separate angles without making clear it was the same scene. And, while the full report precludes drawing conclusions, it is not a document inspired in the fairness of the network news.

Our knowledge of the impact of network news on the national mind is far from complete, but some early returns are available. Again, we have enough information to raise serious questions about its effect on a democratic society. Several years ago Fred Friendly, one of the pioneers of network news, wrote that its missing ingredients were conviction, controversy, and point of view. The networks have compensated with vengeance.

And in the networks' endless pursuit of controversy, we should ask: What is the end value—to enlighten or to profit? What is the end result—to inform or to confuse? How does the ongoing exploration for more action, more excitement, and more drama serve our national search for internal peace and stability?

Gresham's Law seems to be operating in the network news. Bad news drives out good news. The irrational is more controversial than the rational. Concurrence can no longer compete with dissent.

One minute of Eldridge Cleaver is worth 10 minutes of Roy Wilkins. The labor crisis settled at the negotiating table is nothing compared to the confrontation that results in a strike—or better yet, violence along the picket lines.

Normality has become the nemesis of the network news. Now the upshot of all this controversy is that a narrow and distorted picture of America often emerges from the televised news.

A single, dramatic piece of the mosaic becomes in the mind of millions the entire picture. And the American who relies upon television for his news might conclude that the majority of black Americans feel no regard for their country. That violence and lawlessness are the rule rather than the exception on the American campus.

We know that none of these conclusions is true.

Perhaps the place to start looking for a credibility gap is not in the offices of the Government in Washington but in the studios of the networks in New York.

Television may have destroyed the old stereotypes, but has it not created new ones in their places?

What has this passionate pursuit of controversy done to the politics of progress through local compromise essential to the functioning of a democratic society?

The members of Congress or the Senate who follow their principles and philosophy quietly in a spirit of compromise are unknown to many Americans, while the loudest and most extreme dissenters on every issue are known to every man on the street.

How many marches and demonstrations would we have if the marchers did not know that the ever-faithful TV cameras would be there to record their antics for the next news show?

We've heard demands that Senators and Congressmen and judges make known all their financial connections so that the public will know who and what influences their decisions and their votes. Strong arguments can be made for that view.

But when a single commentator or producer, night after night, determines for millions of people how much of each side of a great issue they are going to see and hear, should he not disclose his personal views on the issue as well?

In this search for excitement and controversy, has more than equal time gone to the minority of Americans who specialize in attacking the United States—its institutions and its citizens?

Tonight I've raised questions. I've made no attempt to suggest the answers. The answers must come from the media men. They are challenged to turn their critical powers on themselves, to direct their energy, their talent, and their conviction toward improving the quality and objectivity of news presentation.

They are challenged to structure their own civic ethics to relate to the great responsibilities they hold.

And the people of America are challenged, too, challenged to press for responsible news presentations. The people can let the networks know that they want their news straight and objective. The people can register their complaints on bias through mail to the networks and phone calls to local stations. This is one case where the people must defend themselves; where the citizen, not the

Government, must be the reformer; where the consumer can be the most effective crusader.

By way of conclusion, let me say that every elected leader in the United States depends on these men of the media. Whether what I've said to you tonight will be heard and seen at all by the nation is not my decision, it's not your decision, it's their decision.

In tomorrow's edition of The Des Moines Register, you'll be able to read a news story detailing what I've said tonight. Editorial comment will be reserved for the editorial page, where it belongs.

Should not the same wall of separation exist between news and comment on the nation's networks?

Now, my friends, we'd never trust such power, as I've described, over public opinion in the hands of an elected Government. It's time we questioned it in the hands of a small and unelected elite.

The great networks have dominated America's airwaves for decades. The people are entitled to a full accounting of their stewardship.

Used with permission from *Vital Speeches of the Day* (Southold, NY: City News Publishing Co., 1969).

# Radio-Television News Directors Association Code of Broadcast News Ethics

The responsibility of radio and television journalists is to gather and report information of importance and interest to the public accurately, honestly and impartially.

The members of the Radio-Television News Directors Association accept these standards and will:

1. Strive to present the source or nature of broadcast news material in a way that is balanced, accurate and fair.
   A. They will evaluate information solely on its merits as news, rejecting sensationalism or misleading emphasis in any form.
   B. They will guard against using audio or video material in a way that deceives the audience.
   C. They will not mislead the public by presenting as spontaneous news any material which is staged or rehearsed.
   D. They will identify people by race, creed, nationality or prior status only when it is relevant.
   E. They will clearly label opinion and commentary.
   F. They will promptly acknowledge and correct errors.
2. Strive to conduct themselves in a manner that protects them from conflicts of interest, real or perceived. They will decline gifts or favors which would influence or appear to influence their judgments.
3. Respect the dignity, privacy and well-being of people with whom they deal.
4. Recognize the need to protect confidential sources. They will promise confidentiality only with the intention of keeping that promise.
5. Respect everyone's right to a fair trial.
6. Broadcast the private transmissions of other broadcasters only with permission.
7. Actively encourage observance of this Code by all journalists, whether members of the Radio-Television News Directors Association or not.

Used with permission by the Radio-Television News Directors Association, February 1990.

# Society of Professional Journalists/Sigma Delta Chi Code of Ethics

The Society of Professional Journalists, Sigma Delta Chi, believes the duty of journalists is the serve the truth.

We believe the agencies of mass communication are carriers of public discussion and information, acting on their Constitutional mandate and freedom to learn and report the facts.

We believe in public enlightenment as the forerunner of justice, and in our Constitutional role to seek the truth as part of the public's right to know the truth.

We believe those responsibilities carry obligations that require journalists to perform with intelligence, objectivity, accuracy and fairness.

To these ends, we declare acceptance of the standards of practice here set forth:

### I. Responsibility:
The public's right to know of events of public importance and interest is the overriding mission of the mass media. The purpose of distributing news and enlightened opinion is to serve the general welfare. Journalists who use their professional status as representatives of the public for selfish or other unworthy motives violate a high trust.

### II. Freedom of the press:
Freedom of the press is to be guarded as an inalienable right of people in a free society. It carries with it the freedom and the responsibility to discuss, question and challenge actions and utterances of our government and of our public and private institutions. Journalists uphold the right to speak unpopular opinions and the privilege to agree with the majority.

### III. Ethics:
Journalists must be free of obligation to any interest other than the public's right to know the truth.

1. Gifts, favors, free travel, special treatment or privileges can compromise the integrity of journalists and their employers. Nothing of value should be accepted.
2. Secondary employment, political involvement, holding public office and service in community organizations should be avoided if it compromises the integrity of journalists and their employers. Journalists and their employers should conduct their personal lives in a manner which protects them from conflict of interest, real or apparent. Their responsibilities to the public are paramount. This is the nature of their profession.
3. So-called news communications from private sources should not be published or broadcast without substantiation of their claims to news values.
4. Journalists will seek news that serves the public interest, despite the obstacles. They will make constant efforts to assure that the public's business is conducted in public and that public records are open to public inspection.
5. Journalists acknowledge the newsman's ethic of protecting confidential sources of information.
6. Plagiarism is dishonest and unacceptable.

### IV. Accuracy and objectivity:
Good faith with the public is the foundation of all worthy journalism.

1. Truth is our ultimate goal.
2. Objectivity in reporting the news is another goal which serves as the mark of an experienced professional. It is a standard of performance toward which we strive. We honor those who achieve it.
3. There is no excuse for inaccuracies or lack of thoroughness.
4. Newspaper headlines should be fully warranted by the contents of the articles they accompany. Photographs and telecasts should give an accurate picture of an event and not highlight an event out of context.
5. Sound practice makes clear distinction between news reports and expressions of opinion. News reports should be free of opinion or bias and represent all sides of an issue.
6. Partisanship in editorial comment that knowingly departs from the truth violates the spirit of American journalism.
7. Journalists recognize their responsibility for offering informed analysis, comment, and editorial opinion on public events and issues. They accept the obligation to present such material by individuals whose competence, experience and judgment qualify them for it.
8. Special articles or presentations devoted to advocacy or the writer's own conclusions and interpretations should be labeled as such.

## V. Fair play:

Journalists at all times will show respect for the dignity, privacy, rights and well-being of people encountered in the course of gathering and presenting the news.

1. The news media should not communicate unofficial charges affecting reputation or moral character without giving the accused a chance to reply.
2. The news media must guard against invading a person's right to pri vacy.
3. The media should not pander to morbid curiosity about details of vice and crime.
4. It is the duty of news media to make prompt and complete correction of their errors.
5. Journalists should be accountable to the public for their reports and the public should be encouraged to voice its grievances against the media. Open dialogue with our readers, viewers and listeners should be fos tered.

## VI. Pledge:

Adherence to this code is intended to preserve and strengthen the bond of mutual trust and respect between American journalists and the American people.

The Society shall—by programs of education and other means—encourage individual journalists to adhere to these tenets, and shall encourage journalistic publications and broadcasters to recognize their responsibility to frame codes of ethics in concert with their employees to serve as guidelines in furthering these goals.

Adopted in 1926; revised in 1973, 1984 and 1987. Used with permission by the Society of Professional Journalists/Sigma Delta Chi, February 1990.

# Bibliography

## BOOKS

Aristotle. *Nichomachean Ethics, Book II*. Translated by J.A.K. Thompson. London: George Allen & Unwin, 1953.

Biagi, Shirley. *Media Impact: An Introduction to Mass Media*. Belmont, CA: Wadsworth Publishing Company, 1988.

*Broadcasting/Cable Yearbook '89*. Washington, D.C.: Broadcasting Publications, Inc., 1989.

Cole, Barry G., ed. *Television*. New York: MacMillan and Co., 1970.

Christians, Clifford G., Kim B. Rotzoll, and Mark Fackler. *Media Ethics: Cases and Moral Reasoning, Second Edition*. New York: Longman, Inc., 1987.

Cullen, Maurice R., Jr. *Mass Media & the First Amendment*. Dubuque, IA: Wm. C. Brown Company Publishers, 1981.

Deats, Paul, ed. *Toward a Discipline of Social Ethics*. Boston: Boston University Press, 1972.

Downs, Hugh. *On Camera: My 10,000 Hours on Television*. New York: G.P. Putnam's Sons, 1986.

Fink, Conrad C. *Media Ethics: In the Newsroom and Beyond*. New York: McGraw-Hill Book Company, 1988.

Gowans, Christopher, ed. *Moral Dilemmas*. New York: Oxford University Press, 1987.

Halberstam, David. *The Powers That Be*. New York: Dell Publishing Co., Inc.,1979.

Hulteng, John L. *The Messenger's Motives: Ethical Problems of the News Media, Second Edition*. Englewood Cliffs, NJ: Prentice-Hall, Inc., 1985.

Kultgen, John. *Ethics and Professionalism*. Philadelphia: University of Pennsylvania Press, 1988.

Limberg, Val E. *Mass Media Literacy: An Introduction to Mass Communication*. Dubuque, IA: Kendall/Hunt Publishing Company, 1988.

Matusow, Barbara. *The Evening Stars*. New York: Ballantine Books, 1983.

Meyer, Philip. *Ethical Journalism*. New York: Longman, Inc., 1987.

Minow, Newton. *Equal Time: The Private Broadcaster and the Public Interest*. New York: Atheneum, 1964.

Norback, Craig T. and Peter G., eds. *TV Guide Almanac*. New York: Ballantine Books, 1980.

Pasqua, Thomas M. Jr., James K. Buckalew, Robert E. Rayfield, and James W. Tankard, Jr. *Mass Media in the Information Age.* Englewood Cliffs, NJ: Prentice-Hall, 1990.

Rawls, John. *A Theory of Justice.* Cambridge: Harvard University Belknap Press, 1971.

Shibutani, Tamotsu. *Improvised News: A Sociological Study of Rumors.* Indianapolis-New York: The Bobbs-Merrill Co., 1966.

Sperber, A.M. *Murrow: His Life and Times.* New York: Freundlich, 1986.

## PERIODICALS

Buckro, Casey. "The Code is Intended to Make Us Think Professionally." *Solutions Today for Ethics Problems Tomorrow: A Special Report by the Ethics Committee of the Society of Professional Journalists, 1989,* p. 22.

Costello, Jan. "Exploiting Grief: Restraint & The Right to Know." *Commonweal,* 6 June 1986, pp. 327-329.

Dempsey, John. "INN Facing Challenges in the '90s; Syndex, Clearances are Vexing." *Variety,* 6 September 1989, pp. 67, 76.

_____. "More Mags Will Fly in the Fall; Too Much of a Bad Thing?" *Variety,* 12 April 1989, pp. 79, 98.

Elm, Joanna. "Tonight's Hot Story is Brought to You . . . By You!" *TV Guide,* 24 February 1990, pp. 23-27.

Gay, Verne. "CBS Staged Vietnam Atrocity, Says Book by Army Historian." *Variety,* 19 July 1989, pp. 1-2.

Harris, Paul. "Ethics Issues a Hotter Topic than Usual for RTNDA." *Variety,* 6 September 1989, p. 66.

Ignatieff, Michael. "Is Nothing Sacred? The Ethics of Television." *Daedalus* 114 (1985): 57-77.

Judge, Frank. "Critic Questions News Practices." *The Detroit News,* 18 February 1973, p. 67.

Knight, Bob. "Network Evening News Shares Plummet From 76 to 59 Since 1979-1980." *Variety,* 20 September 1989, p. 124.

McGregor, James. "How ABC's Av Westin Decides What to Show on the Evening News." *Wall St. Journal,* 22 November 1972, p.1.

Montez, Abel. "Computer Age Poses New Obstacles, New Options." *1989-1990 Report of the Society of Professional Journalists,* pp. 3-4.

"Rather Concerned over Afghan Flap." *TV Guide,* 27 January 1990, pp. 42-43.

Smith, Desmond. "TV News Did Not Just Happen—It Had to Invent Itself." *Smithsonian,* June 1989, pp. 74-90.

Stein, Lisa. "Networks, Pentagon in Hot Dispute Over Panama Footage." *TV Guide,* 10 February 1990, pp. 42-43.

Tyler, Ralph. "How Stations Draw Line on Video News Releases." *Variety,* 6 September 1989, pp. 67, 78.

_____. "News Business vs. Show Business is on News Directors' Minds." *Variety*, 6 September 1989, pp. 67, 78.

### TELEVISION PROGRAMS
"The Koppel Report: Television—Revolution in a Box." Produced by Ted Koppel and ABC News, New York, 1989.

"Media Probes: TV News." Produced by WQED-TV, Pittsburgh, 1982.

"Television: TV News—The Power of Pictures." Produced by WNET-TV, New York and KCET-TV, Los Angeles, in cooperation with Granada Television of London, England, 1988.